Electromagnetic Reciprocity in Antenna Theory

Electromagnetic Reciprocity in Antenna Theory

Martin Štumpf

IEEE PRESS

WILEY

Published by John Wiley & Sons, Inc., Hoboken, New Jersey.
Published simultaneously in Canada.

Library of Congress Cataloging-in-Publication Data is available.

ISBN: 978-1-119-46637-6

10 9 8 7 6 5 4 3 2 1

Dedicated to H. A. Lorentz Chair Emeritus Professor Adrianus T. de Hoop
on the occasion of his 90th birthday

Much of the research work included in this book was sponsored by the Czech Science Foundation under Grant 17-05445Y. This research was also partially supported by the Czech Ministry of Education, Youth and Sports under Grant LO1401 (National Sustainability Program). This financial support is gratefully acknowledged.

Special thanks go to Danielle Lacourciere and Mary Hatcher at Wiley-IEEE Press, and to Shikha Pahuja, PM at Thomson Digital, for their professional assistance that helped to publish the book in a timely manner.

Contents

Introduction

Reciprocity theorems are among the most intriguing concepts in the study of wave fields. Since its introduction by H.A. Lorentz in 1896 [33], the concept of reciprocity has become an integral part of (almost) all standard textbooks on electromagnetic (EM) theory (e.g., [27], Sec. 3.8). Its early developments in electrical engineering refer to Rayleigh's (acoustic) reciprocity theorem [32] and are closely related to Kirchhoff's networks [5]. Widespread applications of EM reciprocity were initially obstructed by a somewhat narrow understanding of the concept of reciprocity and its conditions of validity [2,6]. An important step forward in putting EM reciprocity into practical use was the introduction of the concept of reaction [37]. While still limited to the "source-field interaction" only, its illuminating notation has paved the way for the development of novel approximate solutions (e.g., Ref. [8]) and variational computational formulations [26]. The original EM reciprocity with the time-harmonic dependence were extended by Welch, who first presented an EM reciprocity theorem of the time-correlation type applying to homogeneous, isotropic, and lossless media [50] and, subsequently, changed the theorem into its time-convolution type and included conductive losses [51]. A clear distinction between the reciprocity theorem of the time-convolution and time-correlation types has been later given by Bojarski [3] who has stressed the importance of the nature of the time dependence with respect to the radiation condition. His generalized time-domain (TD) reciprocity theorem, however, is restricted to homogeneous and isotropic media, where the EM field representation in terms of the scalar and vector potentials applies. This has been later extended by de Hoop [14], who has introduced the general forms of the EM reciprocity theorems for EM field states in linear, time-invariant, yet arbitrarily inhomogeneous, anisotropic, and dispersive media.

The current understanding of the reciprocity theorem can be briefly stated as follows ([16], Sec. 28.1): "A reciprocity theorem interrelates, in a specific manner, the EM field or wave quantities that characterize two admissible states that could occur in one and the same time-invariant domain of space. Each of the two states can be associated with its own set of time-invariant medium parameters and its own set of source distributions." In this general sense, the reciprocity theorem is no longer understood as a mere statement about the source-observer interchange, but it is rather envisaged as being representative of the quantitative interaction between the observed field quantity (or the EM field to be actually calculated) and an observational action (or a suitable testing state) in any EM measurement (or computational) configuration. Accordingly, the reciprocity theorem encompasses all the formulations of direct/inverse source/scattering problems [16,22,23] and can be thus understood as a truly unifying principle in the wave field modeling [15]. This view is also followed throughout this book, where the EM reciprocity theorem is applied to analyze EM reciprocity properties of multiport antenna systems.

In Chapter 1 we make the reader acquainted with some basic mathematical tools necessary for the reciprocity-based antenna system analysis. In particular, we define the one-sided Laplace transformation and discuss its use in the context of the EM reciprocity theorems of the time-convolution and time-correlation types. Most of the results in the book are then derived under the Laplace transformation, that is, in the complex-frequency domain (FD). Furthermore, the generic (multiport) antenna configuration is described. In particular, we discuss the antenna-system material constitutive parameters and the proper definition of Kirchhoff-type accessible ports.

In Chapter 2 we analyze the conditions that must be placed upon the antenna-system constitutive parameters to get one and the same radiated causal EM wave fields. Since the proof of uniqueness largely relies on Lerch's theorem, this chapter is supplemented with Appendix A, where the theorem along with its proof are given.

In Chapter 3 we provide a special form of the forward-scattering theorem (in the complex-FD) that applies to the analyzed multiport antenna configuration. Its real-FD and time-domain (TD) counterparts are subsequently discussed in the corresponding exercise at the end of the chapter.

In Chapter 4 it is shown that the classic (local) conjugate matching condition has its (global) counterpart expressed in terms of the corresponding scattered and radiated far-field amplitudes. The full equivalence between the two matching conditions is then demonstrated for an important class of small antennas in the accompanying (solved) exercises.

In Chapter 5 we apply the reciprocity theorem of the time-convolution type to one and the same N-port antenna system to find its equivalent Kirchhoff network representations. The corresponding equivalent source generators are specified for the excitation via a known volume-current distribution and via an incident plane wave. The chapter is concluded by a solved exercise, where the relation between the absorption cross section of the load and the antenna gain is rigorously derived for an 1-port antenna system.

The impact of a scatterer on the equivalent Kirchhoff network representations of a multiport antenna system is described in Chapter 6. Both transmitting and receiving situations are studied in detail. In the corresponding exercise, we analyze the contribution of a small raindrop to the equivalent voltage-source strength of a short-wire antenna.

In Chapter 7 the EM reciprocity theorems are applied to analyze the mutual EM coupling between two multiport antenna systems. Here, without loss of generality, the interaction is described in terms of (open-circuited) impedance parameters. Some interesting properties of the transfer-impedance matrix that follow from the application of the time-convolution and time-correlation reciprocity theorems are pointed out.

The EM field transfer between two multiport antenna systems is inevitably affected by the presence of neighboring objects. In Chapter 8 we apply the EM reciprocity theorems to show how the presence of such an object manifests itself in the transfer-impedance matrix. This chapter is further supplemented with an illustrative exercise that results in an efficient formula for lightning-induced EM coupling calculations.

In Chapter 9 we study EM compensation theorems that relate the change of antenna-system EM scattering properties to its radiation characteristics in the corresponding transmitting situation. The obtained compensation relations are next applied to shed some light on the adequacy of the Kirchhoff network representations of a multiport receiving antenna system. The concluding exercises deal with special cases of the derived compensation theorems and with their application to small antennas.

1

Basic Prerequisites

We shall investigate EM radiation, scattering, and reciprocity properties of antenna systems, an example of which is shown in Figure 1.1. To analyze such space–time problems, it is necessary to localize the point in the problem configuration and register the instant at which a given wave field quantity occurs in its evolution following the source excitation. To this end, we will employ the Cartesian reference frame that is defined via its basis vectors $\{\boldsymbol{i}_1, \boldsymbol{i}_2, \boldsymbol{i}_3\}$ and the origin denoted by $\mathcal{O}$. Consequently, the (standard) basis allows specifying the position of an observer via the linear combination:

$$x_1\boldsymbol{i}_1 + x_2\boldsymbol{i}_2 + x_3\boldsymbol{i}_3 \tag{1.1}$$

where $\{x_1, x_2, x_3\}$ are the scalar and real-valued components of the position vector (i.e., spatial coordinates, in short). Such linear combinations of the Cartesian basis vectors can be represented as 1D arrays and will be further represented by boldface symbols. In particular, the position vector will be denoted by $\boldsymbol{x}$ and its components are denoted by x_k for $k \in \{1,2,3\}$. A natural extension in this respect is a Cartesian tensor of rank 2 that can be represented by a 2D array. Such quantities will be denoted by underlined boldface symbols. An example from the category is a quantity denoted by $\underline{\boldsymbol{\eta}}$, for instance, whose components are $\eta_{q,r}$ for $q, r \in \{1,2,3\}$. Consequently, we shall use the following notation for the products between arrays:

$$\boldsymbol{u} \cdot \boldsymbol{v} \quad \Leftrightarrow \quad \sum_{k=1}^{3} u_k v_k \tag{1.2}$$

$$(\underline{\boldsymbol{\zeta}} \cdot \boldsymbol{v}) \cdot \boldsymbol{i}_j \quad \Leftrightarrow \quad \sum_{k=1}^{3} \zeta_{j,k} v_k \text{ for } j \in \{1,2,3\} \tag{1.3}$$

Electromagnetic Reciprocity in Antenna Theory, First Edition. Martin Stumpf.

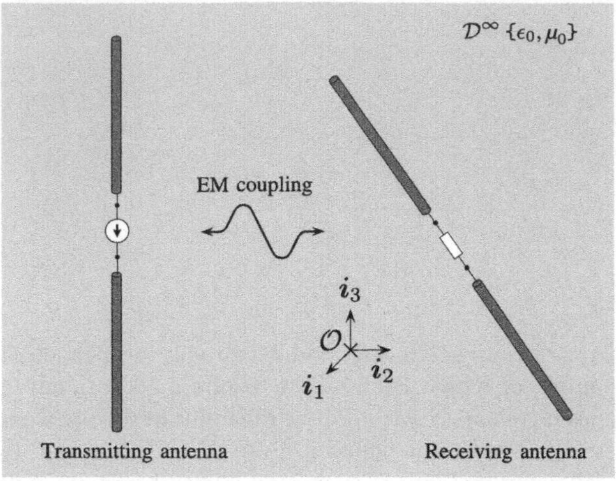

Figure 1.1 Two wire antennas.

$$\boldsymbol{u} \cdot \underline{\boldsymbol{\zeta}} \cdot \boldsymbol{v} \quad \Leftrightarrow \quad \sum_{j=1}^{3}\sum_{k=1}^{3} \zeta_{j,k} u_j v_k \tag{1.4}$$

$$(\boldsymbol{u} \times \boldsymbol{v}) \cdot \boldsymbol{i}_j \quad \Leftrightarrow \quad \sum_{k=1}^{3}\sum_{l=1}^{3} \varepsilon_{j,k,l} u_k v_l \ \text{ for } j \in \{1,2,3\} \tag{1.5}$$

$$\underline{\boldsymbol{\alpha}} \cdot \underline{\boldsymbol{\zeta}} \cdot \underline{\boldsymbol{\beta}} \quad \Leftrightarrow \quad \sum_{p=1}^{3}\sum_{q=1}^{3} \alpha_{k,p} \zeta_{p,q} \beta_{q,n} \ \text{ for } k,n \in \{1,2,3\} \tag{1.6}$$

$$\boldsymbol{u} \times \underline{\boldsymbol{\zeta}} \times \boldsymbol{v} \quad \Leftrightarrow \quad \sum_{r=1}^{N}\sum_{s=1}^{N}\sum_{q=1}^{N}\sum_{j=1}^{N} \varepsilon_{k,r,s} \varepsilon_{l,q,j} \zeta_{s,q} u_r v_j$$
$$\text{for } k,l \in \{1,2,3\} \tag{1.7}$$

where $\varepsilon_{j,k,l}$ is the Levi-Civita tensor (= the completely antisymmetrical unit tensor of rank 3) defined as $\varepsilon_{j,k,l} = 1$ for $\{j,k,l\}$ = even permutation of $\{1,2,3\}$, $\varepsilon_{j,k,l} = -1$ for $\{j,k,l\}$ = odd permutation of $\{1,2,3\}$, and $\varepsilon_{j,k,l} = 0$ in all other cases ([16], Sec. A.7). Finally, for a tensor of rank 2, say $\underline{\boldsymbol{\zeta}}$, the tensor transpose is denoted by $\underline{\boldsymbol{\zeta}}^{\mathcal{T}}$ and its components are found according to

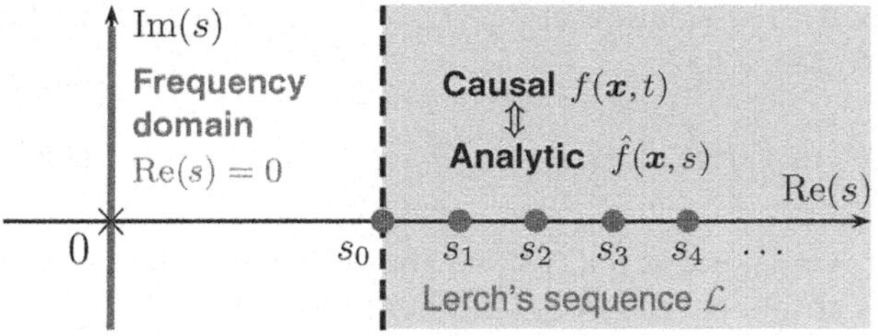

Figure 1.2 Complex frequency plane.

$$\zeta^{T}_{k,l} = \zeta_{l,k} \tag{1.8}$$

for all $k, l \in \{1, 2, 3\}$.

The time coordinate is real-valued and will be denoted by t. Since we assume that the sources generating the EM wave fields are switched on at the origin $t = 0$, we will analyze the excited EM field quantities in $\{t \in \mathbb{R}; t > 0\}$ only, which is possible in virtue of the universal property of causality.

1.1 Laplace Transformation

The one-sided Laplace transformation of a wave quantity $f(\boldsymbol{x}, t)$ is defined by the following integral:

$$\hat{f}(\boldsymbol{x}, s) = \mathsf{L}\{f(\boldsymbol{x}, t)\} = \int_{t=0}^{\infty} \exp(-st) f(\boldsymbol{x}, t) \mathrm{d}t \tag{1.9}$$

In the definition of Eq. (1.9), we shall limit ourselves to physical wave quantities that are bounded such that $f(\boldsymbol{x}, t) = O[\exp(s_0 t)]$[1] with $s_0 \in \mathbb{R}$ being its exponential order . For such functions, the Laplace integral converges if s is either real-valued and positive with $s > s_0$ or complex-valued with $\mathrm{Re}(s) > s_0$. Accordingly, the right-half plane $\mathrm{Re}(s) > s_0$ is the domain of regularity of the causal wave quantity (see Figure 1.2). If Eq. (1.9) is viewed as an integral equation to be solved for the unknown function $f(\boldsymbol{x}, t)$, a natural question as to its uniqueness arises. The question has been convincingly settled by Lerch who proved that the image function known along the Lerch sequence $\mathcal{L} = \{s \in \mathbb{R}, s = s_0 + nh, h > 0, n = 1, 2, \ldots\}$ results in one and the same causal original function. Since single-point discontinuities are of no practical importance, the ambiguity brought about by null functions may be ignored for our

1 By $\phi(x) = O[h(x)]$, we mean that $|\phi(x)| < A|h(x)|$ for $\{A \in \mathbb{R}; A > 0\}$. In particular, $\phi(x) = O(1)$ represents a bounded function.

purposes (see Appendix A). The solution of Eq. (1.9) can be expressed via the Bromwich inversion integral:

$$f(\boldsymbol{x},t) = \frac{1}{2\pi \mathrm{i}} \int_{s\in \mathcal{B}_r} \exp(st)\hat{f}(\boldsymbol{x},s)\mathrm{d}s \tag{1.10}$$

where $\mathcal{B}_r = \{s \in \mathbb{C}; \mathrm{Re}(s) = s_0, -\infty < \mathrm{Im}(s) < \infty\}$ and we have tacitly assumed that Eq. (1.9) converges absolutely at $s = s_0$ (and hence for all $\mathrm{Re}(s) > s_0$). A special, yet frequently used, case arises for (bounded) functions of exponential order $s_0 = 0$ for which the Bromwich integral (1.10) can be rewritten using the limit $s = \delta + \mathrm{i}\omega$ as $\delta \downarrow 0$, where ω is the (real-valued) angular frequency. Under this limit, Eqs. (1.9) and (1.10) express the Fourier transform of a causal wave quantity. For more details about the Laplace transformation, we refer the reader to Refs. [16,20,47,52].

1.2 Time Convolution

The time convolution between the *causal* wave quantities $f(\boldsymbol{x},t)$ and $g(\boldsymbol{x},t)$ is defined as

$$[f * g](\boldsymbol{x},t) = \int_{\tau\in\mathbb{R}} f(\boldsymbol{x},\tau)g(\boldsymbol{x},t-\tau)\mathrm{d}\tau \tag{1.11}$$

An important property of the time-convolution operator is its commutativity allowing to rewrite the latter as

$$\begin{aligned}[g * f](\boldsymbol{x},t) &= \int_{\tau\in\mathbb{R}} f(\boldsymbol{x},t-\tau)g(\boldsymbol{x},\tau)\mathrm{d}\tau \\ &= \int_{\tau\in\mathbb{R}} f(\boldsymbol{x},\tau)g(\boldsymbol{x},t-\tau)\mathrm{d}\tau = [f * g](\boldsymbol{x},t)\end{aligned} \tag{1.12}$$

Applying the Laplace transformation to Eq. (1.11) yields

$$\mathsf{L}\{[f * g](\boldsymbol{x},t)\} = \hat{f}(\boldsymbol{x},s)\hat{g}(\boldsymbol{x},s) \tag{1.13}$$

For Eq. (1.13) to make any sense, there must be at least one value of s for which the Laplace integrals for $\hat{f}(\boldsymbol{x},s)$ and $\hat{g}(\boldsymbol{x},s)$ do converge simultaneously. If $f(\boldsymbol{x},t) = O[\exp(s_0 t)]$ and $g(\boldsymbol{x},t) = O[\exp(\sigma_0 t)]$, the region of convergence for Eq. (1.13) is found in the domain extending to the right of $\max\{s_0, \sigma_0\}$ (see Figure 1.3 for $\sigma_0 > s_0$). Consequently, the right half-plane of the complex frequency plane $\mathrm{Re}(s) > \max\{s_0, \sigma_0\}$ is the domain of regularity of the Laplace-transformed time convolution given in Eq. (1.13).

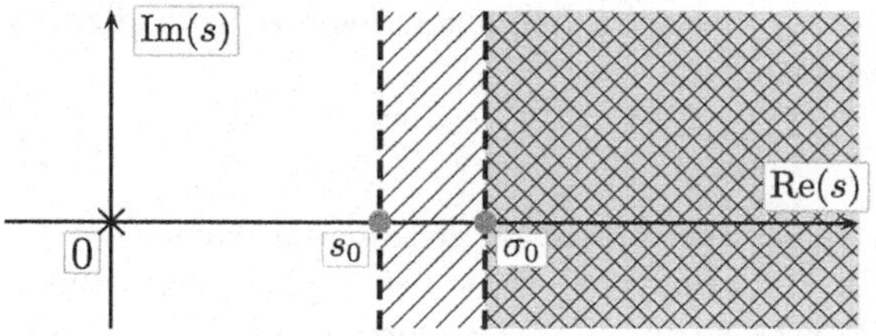

Figure 1.3 Region of convergence of $\hat{f}(\boldsymbol{x}, s)\hat{g}(\boldsymbol{x}, s)$ that is found as the intersection of $\mathrm{Re}(s) > s_0$ and $\mathrm{Re}(s) > \sigma_0$.

1.3 Time Correlation

The time correlation between the *causal* wave quantities $f(\boldsymbol{x}, t)$ and $g(\boldsymbol{x}, t)$ is defined as

$$[f \star g](\boldsymbol{x}, t) = \int_{\tau \in \mathbb{R}} f(\boldsymbol{x}, \tau)g(\boldsymbol{x}, \tau - t)\mathrm{d}\tau \tag{1.14}$$

In contrast to the time-convolution operator, the time-correlation one is not commutative. Indeed, Eq. (1.14) can be rewritten as

$$[f \star g](\boldsymbol{x}, t) = \int_{\tau \in \mathbb{R}} f(\boldsymbol{x}, \tau + t)g(\boldsymbol{x}, \tau)\mathrm{d}\tau \tag{1.15}$$

which implies that

$$[f \star g](\boldsymbol{x}, t) = [g \star f](\boldsymbol{x}, -t) \tag{1.16}$$

Also, by inspection of Eqs. (1.11) and (1.14), we may write

$$[f \star g](\boldsymbol{x}, t) = [f * g^{\circledast}](\boldsymbol{x}, t)\mathrm{d}\tau \tag{1.17}$$

where superscript $\circledast$ applied to a *space–time* field quantity represents the time-reversal operator, that is,

$$g^{\circledast}(\boldsymbol{x}, t) = g(\boldsymbol{x}, -t) \tag{1.18}$$

Obviously, the time-reversed causal wave quantity has its support in $\{t \in \mathbb{R}; t < 0\}$, which in accordance with Eq. (1.9) implies that the domain of convergence of its image function extends over a left half of the complex s-plane. Combination of Eqs. (1.17) and (1.18) with (1.13) and (1.9) then implies that the Laplace transformation of Eq. (1.14) can be, yet formally only, written as

$$\mathsf{L}\{[f \star g](\boldsymbol{x}, t)\} = \hat{f}(\boldsymbol{x}, s)\hat{g}^{\circledast}(\boldsymbol{x}, s) \tag{1.19}$$

where superscript ⊛ applied to a *complex frequency domain* field quantity represents the following operation:

$$\hat{g}^{\circledast}(\boldsymbol{x}, s) = \hat{g}(\boldsymbol{x}, -s) \tag{1.20}$$

For Eq. (1.19) to make any sense, there must be at least one value of s for which the Laplace integrals for $\hat{f}(\boldsymbol{x}, s)$ and $\hat{g}(\boldsymbol{x}, -s)$ do converge simultaneously. Again, assuming $f(\boldsymbol{x}, t) = O[\exp(s_0 t)]$ and $g(\boldsymbol{x}, t) = O[\exp(\sigma_0 t)]$, the region of convergence for Eq. (1.19) is found as the intersection of the domains of convergence corresponding to $\hat{f}(\boldsymbol{x}, s)$ and $\hat{g}(\boldsymbol{x}, -s)$. Hence, this region of convergence is at most a strip of the complex frequency plane bounded by two verticals $\{s_0 < \mathrm{Re}(s) < \sigma_0\}$ (see Figure 1.4). In this respect it should be emphasized that the vast majority of wave field quantities we will deal with are bounded functions of exponential order $s_0 = 0$. For such a class of functions, Eq. (1.19) makes a sense along the imaginary axis in the complex frequency plane only, that is, in the limiting real-frequency domain for $s = \delta + \mathrm{i}\omega$ with $\delta \downarrow 0$ and $\omega \in \mathbb{R}$. Note that in such a case, superscript ⊛ in Eq. (1.20) has the meaning of complex conjugate, that is, $\hat{g}^{\circledast}(\boldsymbol{x}, \mathrm{i}\omega) = \hat{g}(\boldsymbol{x}, -\mathrm{i}\omega)$ since $g(\boldsymbol{x}, t)$ is real-valued.

1.4 EM Reciprocity Theorems

In this section, a brief review concerning EM reciprocity theorems is given. In accordance with Sec. 28 of Ref. [16], we shall distinguish between the reciprocity theorems of the time-convolution and time-correlation types. The reciprocity relations will be given in the complex-frequency domain.

To find the reciprocity theorems in their generic form, we shall interrelate two states of EM fields, say A and B, that are governed by the EM field (Maxwell) equations ([16], Sec. 24.4):

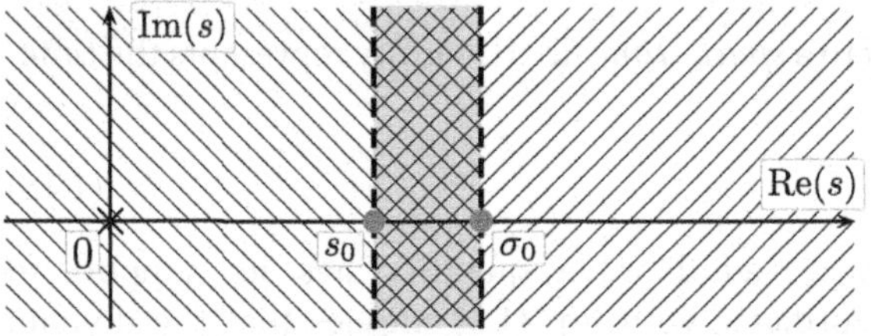

Figure 1.4 Region of convergence of $\hat{f}(\boldsymbol{x}, s)\hat{g}(\boldsymbol{x}, -s)$ that is found as the intersection of $\mathrm{Re}(s) > s_0$ and $\mathrm{Re}(s) < \sigma_0$.

$$\nabla \times \hat{\boldsymbol{H}}^{\mathrm{A,B}} - \underline{\hat{\eta}}^{\mathrm{A,B}} \cdot \hat{\boldsymbol{E}}^{\mathrm{A,B}} = \hat{\boldsymbol{J}}^{\mathrm{A,B}} \tag{1.21}$$

$$\nabla \times \hat{\boldsymbol{E}}^{\mathrm{A,B}} + \underline{\hat{\zeta}}^{\mathrm{A,B}} \cdot \hat{\boldsymbol{H}}^{\mathrm{A,B}} = -\hat{\boldsymbol{K}}^{\mathrm{A,B}} \tag{1.22}$$

for $\boldsymbol{x} \in \mathcal{D}$, where

- $\hat{\boldsymbol{E}}^{\mathrm{A,B}}$ = electric field strength (V/m),
- $\hat{\boldsymbol{H}}^{\mathrm{A,B}}$ = magnetic field strength (A/m),
- $\hat{\boldsymbol{J}}^{\mathrm{A,B}}$ = electric current volume density (A/m^2),
- $\hat{\boldsymbol{K}}^{\mathrm{A,B}}$ = magnetic current volume density (V/m^2),
- $\underline{\hat{\eta}}^{\mathrm{A,B}}$ = transverse admittance per length of the medium (S/m),
- $\underline{\hat{\zeta}}^{\mathrm{A,B}}$ = longitudinal impedance per length of the medium (Ω/m).

Equations (1.21) and (1.22) are in $\mathcal{D}$ further supplemented with the constitutive relations:

$$\underline{\hat{\eta}}^{\mathrm{A,B}}(\boldsymbol{x}, s) = \underline{\hat{\sigma}}^{\mathrm{A,B}}(\boldsymbol{x}, s) + s\underline{\hat{\epsilon}}^{\mathrm{A,B}}(\boldsymbol{x}, s) \tag{1.23}$$

$$\underline{\hat{\zeta}}^{\mathrm{A,B}}(\boldsymbol{x}, s) = s\underline{\hat{\mu}}^{\mathrm{A,B}}(\boldsymbol{x}, s) \tag{1.24}$$

where

- $\underline{\hat{\sigma}}^{\mathrm{A,B}}$ = electric conductivity (S/m),
- $\underline{\hat{\epsilon}}^{\mathrm{A,B}}$ = electric permittivity (F/m), and
- $\underline{\hat{\mu}}^{\mathrm{A,B}}$ = magnetic permeability (H/m).

The medium described by such constitutive relations can be inhomogeneous, anisotropic, and dispersive in its EM behavior. A special yet useful case of Eqs. (1.23) and (1.24) describes an instantaneously reacting (dispersion-free) medium:

$$\underline{\hat{\eta}}^{\mathrm{A,B}}(\boldsymbol{x}, s) = s\underline{\epsilon}^{\mathrm{A,B}}(\boldsymbol{x}) \tag{1.25}$$

$$\underline{\hat{\zeta}}^{\mathrm{A,B}}(\boldsymbol{x}, s) = s\underline{\mu}^{\mathrm{A,B}}(\boldsymbol{x}) \tag{1.26}$$

For isotropic materials the latter relations further simplify to

$$\underline{\hat{\eta}}^{\mathrm{A,B}}(\boldsymbol{x}, s) = s\epsilon^{\mathrm{A,B}}(\boldsymbol{x})\underline{\boldsymbol{I}} \tag{1.27}$$

$$\underline{\hat{\zeta}}^{\mathrm{A,B}}(\boldsymbol{x}, s) = s\mu^{\mathrm{A,B}}(\boldsymbol{x})\underline{\boldsymbol{I}} \tag{1.28}$$

where $\underline{\boldsymbol{I}}$ denotes the 3×3 identity matrix.

1.4.1 Reciprocity Theorem of the Time-Convolution Type

The reciprocity theorem of the time-convolution type is constructed from the following local interaction quantity:

$$\boldsymbol{\nabla}\cdot\left[\hat{\boldsymbol{E}}^{\mathrm{A}}(\boldsymbol{x},s)\times\hat{\boldsymbol{H}}^{\mathrm{B}}(\boldsymbol{x},s)-\hat{\boldsymbol{E}}^{\mathrm{B}}(\boldsymbol{x},s)\times\hat{\boldsymbol{H}}^{\mathrm{A}}(\boldsymbol{x},s)\right] \tag{1.29}$$

for $\boldsymbol{x}\in\mathcal{D}$, that is with the help of Eqs. (1.21) and (1.22) and Gauss' theorem rearranged to its global form ([16], Eq. (28.4-7)):

$$\begin{aligned}
&\int_{\boldsymbol{x}\in\partial\mathcal{D}}\left(\hat{\boldsymbol{E}}^{\mathrm{A}}\times\hat{\boldsymbol{H}}^{\mathrm{B}}-\hat{\boldsymbol{E}}^{\mathrm{B}}\times\hat{\boldsymbol{H}}^{\mathrm{A}}\right)\cdot\boldsymbol{\nu}\,\mathrm{d}A\\
&\quad=\int_{\boldsymbol{x}\in\mathcal{D}}\Big\{\hat{\boldsymbol{H}}^{\mathrm{A}}\cdot\left[\underline{\hat{\zeta}}^{\mathrm{B}}-(\underline{\hat{\zeta}}^{\mathrm{A}})^{\mathcal{T}}\right]\cdot\hat{\boldsymbol{H}}^{\mathrm{B}}\\
&\qquad-\hat{\boldsymbol{E}}^{\mathrm{A}}\cdot\left[\underline{\hat{\eta}}^{\mathrm{B}}-(\underline{\hat{\eta}}^{\mathrm{A}})^{\mathcal{T}}\right]\cdot\hat{\boldsymbol{E}}^{\mathrm{B}}\Big\}\mathrm{d}V\\
&\qquad+\int_{\boldsymbol{x}\in\mathcal{D}}\Big(\hat{\boldsymbol{J}}^{\mathrm{A}}\cdot\hat{\boldsymbol{E}}^{\mathrm{B}}-\hat{\boldsymbol{K}}^{\mathrm{A}}\cdot\hat{\boldsymbol{H}}^{\mathrm{B}}\\
&\qquad-\hat{\boldsymbol{J}}^{\mathrm{B}}\cdot\hat{\boldsymbol{E}}^{\mathrm{A}}+\hat{\boldsymbol{K}}^{\mathrm{B}}\cdot\hat{\boldsymbol{H}}^{\mathrm{A}}\Big)\mathrm{d}V
\end{aligned} \tag{1.30}$$

The first integral on the right-hand side represents the interaction of the field and material states. As this interaction is proportional to the contrast in the EM constitutive properties of the two states, this integral vanishes whenever

$$\underline{\hat{\eta}}^{\mathrm{B}}(\boldsymbol{x},s)=(\underline{\hat{\eta}}^{\mathrm{A}})^{\mathcal{T}}(\boldsymbol{x},s) \tag{1.31}$$

$$\underline{\hat{\zeta}}^{\mathrm{B}}(\boldsymbol{x},s)=(\underline{\hat{\zeta}}^{\mathrm{A}})^{\mathcal{T}}(\boldsymbol{x},s) \tag{1.32}$$

for all $\boldsymbol{x}\in\mathcal{D}$. In such a case, the media are denoted as each other's adjoint. Note in this respect that the latter conditions boil down to $\underline{\boldsymbol{\epsilon}}^{\mathrm{B}}(\boldsymbol{x})=(\underline{\boldsymbol{\epsilon}}^{\mathrm{A}})^{\mathcal{T}}(\boldsymbol{x})$ and $\underline{\boldsymbol{\mu}}^{\mathrm{B}}(\boldsymbol{x})=(\underline{\boldsymbol{\mu}}^{\mathrm{A}})^{\mathcal{T}}(\boldsymbol{x})$ throughout $\mathcal{D}$ for instantaneously reacting media described by Eqs. (1.25)–(1.26) and to $\epsilon^{\mathrm{B}}(\boldsymbol{x})=\epsilon^{\mathrm{A}}(\boldsymbol{x})$ and $\mu^{\mathrm{B}}(\boldsymbol{x})=\mu^{\mathrm{A}}(\boldsymbol{x})$ throughout $\mathcal{D}$ for instantaneously reacting, isotropic media described via Eqs. (1.27) and (1.28). Finally, the key ingredients constituting the time-convolution field–field, material–field, and source–field interactions in domain $\mathcal{D}$ are summarized in Table 1.1. Constitutive relations (1.25) and (1.26) describing the EM behavior of an instantaneously reacting, anisotropic, and inhomogeneous medium lead to

Table 1.1 Application of the reciprocity theorem of the time-convolution type.

Domain $\mathcal{D}$		
Time-convolution	State (A)	State (B)
Source	$\{\hat{\boldsymbol{J}}^{\mathrm{A}}, \hat{\boldsymbol{K}}^{\mathrm{A}}\}$	$\{\hat{\boldsymbol{J}}^{\mathrm{B}}, \hat{\boldsymbol{K}}^{\mathrm{B}}\}$
Field	$\{\hat{\boldsymbol{E}}^{\mathrm{A}}, \hat{\boldsymbol{H}}^{\mathrm{A}}\}$	$\{\hat{\boldsymbol{E}}^{\mathrm{B}}, \hat{\boldsymbol{H}}^{\mathrm{B}}\}$
Material	$\{\hat{\underline{\eta}}^{\mathrm{A}}, \hat{\underline{\zeta}}^{\mathrm{A}}\}$	$\{\hat{\underline{\eta}}^{\mathrm{B}}, \hat{\underline{\zeta}}^{\mathrm{B}}\}$

$$\begin{aligned}
&\int_{\boldsymbol{x}\in\partial\mathcal{D}} \left(\hat{\boldsymbol{E}}^{\mathrm{A}} \times \hat{\boldsymbol{H}}^{\mathrm{B}} - \hat{\boldsymbol{E}}^{\mathrm{B}} \times \hat{\boldsymbol{H}}^{\mathrm{A}}\right) \cdot \boldsymbol{\nu}\,\mathrm{d}A \\
&= s\int_{\boldsymbol{x}\in\mathcal{D}} \Big\{\hat{\boldsymbol{H}}^{\mathrm{A}} \cdot \left[\underline{\mu}^{\mathrm{B}} - (\underline{\mu}^{\mathrm{A}})^{\mathcal{T}}\right] \cdot \hat{\boldsymbol{H}}^{\mathrm{B}} \\
&\qquad - \hat{\boldsymbol{E}}^{\mathrm{A}} \cdot \left[\underline{\epsilon}^{\mathrm{B}} - (\underline{\epsilon}^{\mathrm{A}})^{\mathcal{T}}\right] \cdot \hat{\boldsymbol{E}}^{\mathrm{B}}\Big\}\mathrm{d}V \\
&\qquad + \int_{\boldsymbol{x}\in\mathcal{D}} \Big(\hat{\boldsymbol{J}}^{\mathrm{A}} \cdot \hat{\boldsymbol{E}}^{\mathrm{B}} - \hat{\boldsymbol{K}}^{\mathrm{A}} \cdot \hat{\boldsymbol{H}}^{\mathrm{B}} \\
&\qquad\qquad - \hat{\boldsymbol{J}}^{\mathrm{B}} \cdot \hat{\boldsymbol{E}}^{\mathrm{A}} + \hat{\boldsymbol{K}}^{\mathrm{B}} \cdot \hat{\boldsymbol{H}}^{\mathrm{A}}\Big)\mathrm{d}V
\end{aligned} \tag{1.33}$$

which can be further simplified for an isotropic medium to

$$\begin{aligned}
&\int_{\boldsymbol{x}\in\partial\mathcal{D}} \left(\hat{\boldsymbol{E}}^{\mathrm{A}} \times \hat{\boldsymbol{H}}^{\mathrm{B}} - \hat{\boldsymbol{E}}^{\mathrm{B}} \times \hat{\boldsymbol{H}}^{\mathrm{A}}\right) \cdot \boldsymbol{\nu}\,\mathrm{d}A \\
&= s\int_{\boldsymbol{x}\in\mathcal{D}} \left[\left(\mu^{\mathrm{B}} - \mu^{\mathrm{A}}\right)\hat{\boldsymbol{H}}^{\mathrm{A}} \cdot \hat{\boldsymbol{H}}^{\mathrm{B}} - \left(\epsilon^{\mathrm{B}} - \epsilon^{\mathrm{A}}\right)\hat{\boldsymbol{E}}^{\mathrm{A}} \cdot \hat{\boldsymbol{E}}^{\mathrm{B}}\right]\mathrm{d}V \\
&+ \int_{\boldsymbol{x}\in\mathcal{D}} \left(\hat{\boldsymbol{J}}^{\mathrm{A}} \cdot \hat{\boldsymbol{E}}^{\mathrm{B}} - \hat{\boldsymbol{K}}^{\mathrm{A}} \cdot \hat{\boldsymbol{H}}^{\mathrm{B}} - \hat{\boldsymbol{J}}^{\mathrm{B}} \cdot \hat{\boldsymbol{E}}^{\mathrm{A}} + \hat{\boldsymbol{K}}^{\mathrm{B}} \cdot \hat{\boldsymbol{H}}^{\mathrm{A}}\right)\mathrm{d}V
\end{aligned} \tag{1.34}$$

in accordance with Eqs. (1.27) and (1.28). The literature on the subject is frequently limited to the simplest form of the reciprocity theorem of the time-convolution type (e.g., Refs. [27], Sec. 3.8; [29], Sec. 2.11; and [30], Sec. 5.5).

1.4.2 Reciprocity Theorem of the Time-Correlation Type

The reciprocity theorem of the time-correlation type is constructed from the following local interaction quantity.

$$\begin{aligned}
&\boldsymbol{\nabla} \cdot \left[\hat{\boldsymbol{E}}^{\mathrm{A}}(\boldsymbol{x},s) \times \hat{\boldsymbol{H}}^{\mathrm{B}\circledast}(\boldsymbol{x},s) + \hat{\boldsymbol{E}}^{\mathrm{B}\circledast}(\boldsymbol{x},s) \times \hat{\boldsymbol{H}}^{\mathrm{A}}(\boldsymbol{x},s)\right] \\
&= \boldsymbol{\nabla} \cdot \left[\hat{\boldsymbol{E}}^{\mathrm{A}}(\boldsymbol{x},s) \times \hat{\boldsymbol{H}}^{\mathrm{B}}(\boldsymbol{x},-s) + \hat{\boldsymbol{E}}^{\mathrm{B}}(\boldsymbol{x},-s) \times \hat{\boldsymbol{H}}^{\mathrm{A}}(\boldsymbol{x},s)\right]
\end{aligned} \tag{1.35}$$

for $\boldsymbol{x} \in D$, that is with the help of Eqs. (1.21) and (1.22) and Gauss' theorem rearranged to its global form ([16], Eq. (28.5-7)):

$$\begin{aligned}
&\int_{\boldsymbol{x}\in\partial D} \left(\hat{\boldsymbol{E}}^{\mathrm{A}} \times \hat{\boldsymbol{H}}^{\mathrm{B}\circledast} + \hat{\boldsymbol{E}}^{\mathrm{B}\circledast} \times \hat{\boldsymbol{H}}^{\mathrm{A}}\right) \cdot \boldsymbol{\nu}\,\mathrm{d}A \\
&\quad = -\int_{\boldsymbol{x}\in D} \Big\{ \hat{\boldsymbol{H}}^{\mathrm{A}} \cdot \left[\underline{\hat{\zeta}}^{\mathrm{B}\circledast} + (\underline{\hat{\zeta}}^{\mathrm{A}})^{\mathcal{T}}\right] \cdot \hat{\boldsymbol{H}}^{\mathrm{B}\circledast} \\
&\qquad + \hat{\boldsymbol{E}}^{\mathrm{A}} \cdot \left[\underline{\hat{\eta}}^{\mathrm{B}\circledast} + (\underline{\hat{\eta}}^{\mathrm{A}})^{\mathcal{T}}\right] \cdot \hat{\boldsymbol{E}}^{\mathrm{B}\circledast} \Big\} \mathrm{d}V \\
&\qquad - \int_{\boldsymbol{x}\in D} \Big(\hat{\boldsymbol{J}}^{\mathrm{A}} \cdot \hat{\boldsymbol{E}}^{\mathrm{B}\circledast} + \hat{\boldsymbol{K}}^{\mathrm{A}} \cdot \hat{\boldsymbol{H}}^{\mathrm{B}\circledast} \\
&\qquad + \hat{\boldsymbol{J}}^{\mathrm{B}\circledast} \cdot \hat{\boldsymbol{E}}^{\mathrm{A}} + \hat{\boldsymbol{K}}^{\mathrm{B}\circledast} \cdot \hat{\boldsymbol{H}}^{\mathrm{A}} \Big) \mathrm{d}V
\end{aligned} \tag{1.36}$$

The first integral on the right hand-side represents the interaction of the field and material states. As this interaction is proportional to the contrast in the EM constitutive properties of the two states, this integral vanishes whenever

$$\underline{\hat{\eta}}^{\mathrm{B}}(\boldsymbol{x}, -s) = -(\underline{\hat{\eta}}^{\mathrm{A}})^{\mathcal{T}}(\boldsymbol{x}, s) \tag{1.37}$$

$$\underline{\hat{\zeta}}^{\mathrm{B}}(\boldsymbol{x}, -s) = -(\underline{\hat{\zeta}}^{\mathrm{A}})^{\mathcal{T}}(\boldsymbol{x}, s) \tag{1.38}$$

for all $\boldsymbol{x} \in D$. In such a case, the media are denoted as each other's time-reverse adjoint. Note in this respect that for instantaneously-reacting media described by Eqs. (1.25)–(1.28), the latter conditions have the same form as the one applying to adjoint media (see Eqs. (1.39) and (1.40)). Finally, the key ingredients constituting the time-correlation field-field, material–field, and source–field interactions in domain D are summarized in Table 1.2.

Table 1.2 Application of the reciprocity theorem of the time-correlation type.

Domain D		
Time-correlation	State (A)	State (B)
Source	$\{\hat{\boldsymbol{J}}^{\mathrm{A}}, \hat{\boldsymbol{K}}^{\mathrm{A}}\}$	$\{\hat{\boldsymbol{J}}^{\mathrm{B}}, \hat{\boldsymbol{K}}^{\mathrm{B}}\}$
Field	$\{\hat{\boldsymbol{E}}^{\mathrm{A}}, \hat{\boldsymbol{H}}^{\mathrm{A}}\}$	$\{\hat{\boldsymbol{E}}^{\mathrm{B}}, \hat{\boldsymbol{H}}^{\mathrm{B}}\}$
Material	$\{\underline{\hat{\eta}}^{\mathrm{A}}, \underline{\hat{\zeta}}^{\mathrm{A}}\}$	$\{\underline{\hat{\eta}}^{\mathrm{B}}, \underline{\hat{\zeta}}^{\mathrm{B}}\}$

Constitutive relations (1.25) and (1.26) describing the EM behavior of an instantaneously reacting, anisotropic, and inhomogeneous medium leads to

$$\begin{aligned}
&\int_{\boldsymbol{x}\in\partial\mathcal{D}} \left(\hat{\boldsymbol{E}}^{\mathrm{A}} \times \hat{\boldsymbol{H}}^{\mathrm{B}\circledast} + \hat{\boldsymbol{E}}^{\mathrm{B}\circledast} \times \hat{\boldsymbol{H}}^{\mathrm{A}}\right) \cdot \boldsymbol{\nu}\,\mathrm{d}A \\
&\quad = s \int_{\boldsymbol{x}\in\mathcal{D}} \Big\{ \hat{\boldsymbol{H}}^{\mathrm{A}} \cdot \left[\underline{\mu}^{\mathrm{B}} - (\underline{\mu}^{\mathrm{A}})^{\mathcal{T}}\right] \cdot \hat{\boldsymbol{H}}^{\mathrm{B}\circledast} \\
&\qquad + \hat{\boldsymbol{E}}^{\mathrm{A}} \cdot \left[\underline{\epsilon}^{\mathrm{B}} - (\underline{\epsilon}^{\mathrm{A}})^{\mathcal{T}}\right] \cdot \hat{\boldsymbol{E}}^{\mathrm{B}\circledast} \Big\}\mathrm{d}V \\
&\qquad - \int_{\boldsymbol{x}\in\mathcal{D}} \Big(\hat{\boldsymbol{J}}^{\mathrm{A}} \cdot \hat{\boldsymbol{E}}^{\mathrm{B}\circledast} + \hat{\boldsymbol{K}}^{\mathrm{A}} \cdot \hat{\boldsymbol{H}}^{\mathrm{B}\circledast} \\
&\qquad + \hat{\boldsymbol{J}}^{\mathrm{B}\circledast} \cdot \hat{\boldsymbol{E}}^{\mathrm{A}} + \hat{\boldsymbol{K}}^{\mathrm{B}\circledast} \cdot \hat{\boldsymbol{H}}^{\mathrm{A}}\Big)\mathrm{d}V
\end{aligned} \tag{1.39}$$

which can be further simplified for an isotropic medium to

$$\begin{aligned}
&\int_{\boldsymbol{x}\in\partial\mathcal{D}} \left(\hat{\boldsymbol{E}}^{\mathrm{A}} \times \hat{\boldsymbol{H}}^{\mathrm{B}\circledast} + \hat{\boldsymbol{E}}^{\mathrm{B}\circledast} \times \hat{\boldsymbol{H}}^{\mathrm{A}}\right) \cdot \boldsymbol{\nu}\,\mathrm{d}A \\
&= s \int_{\boldsymbol{x}\in\mathcal{D}} \left\{ \left(\mu^{\mathrm{B}} - \mu^{\mathrm{A}}\right) \hat{\boldsymbol{H}}^{\mathrm{A}} \cdot \hat{\boldsymbol{H}}^{\mathrm{B}\circledast} + \left(\epsilon^{\mathrm{B}} - \epsilon^{\mathrm{A}}\right) \hat{\boldsymbol{E}}^{\mathrm{A}} \cdot \hat{\boldsymbol{E}}^{\mathrm{B}\circledast} \right\}\mathrm{d}V \\
&- \int_{\boldsymbol{x}\in\mathcal{D}} \left(\hat{\boldsymbol{J}}^{\mathrm{A}} \cdot \hat{\boldsymbol{E}}^{\mathrm{B}\circledast} + \hat{\boldsymbol{K}}^{\mathrm{A}} \cdot \hat{\boldsymbol{H}}^{\mathrm{B}\circledast} + \hat{\boldsymbol{J}}^{\mathrm{B}\circledast} \cdot \hat{\boldsymbol{E}}^{\mathrm{A}} + \hat{\boldsymbol{K}}^{\mathrm{B}\circledast} \cdot \hat{\boldsymbol{H}}^{\mathrm{A}}\right)\mathrm{d}V
\end{aligned} \tag{1.40}$$

in accordance with Eqs. (1.27) and (1.28). The importance of the nature of the temporal behavior of the interacting wave field quantities has been stressed by Bojarski [3] who has introduced the time-convolution-and time-correlation-type reciprocity theorems applying to homogeneous, isotropic, and lossless media. This has been later clearly unified by de Hoop [14] who has further generalized the reciprocity theorems by including general inhomogeneous, anisotropic, and dispersive media.

1.4.3 Application of the Reciprocity Theorems to an Unbounded Domain

Whenever a reciprocity theorem is applied to an unbounded domain exterior to an antenna system, the surface integrals that appear in Eqs. (1.30) and (1.36) will be carried out over the outer bounding surface $\partial\mathcal{D}^{\Delta}$ under the limit $\Delta \to \infty$ (see Figure 1.5). To evaluate this contribution for *causal* EM field states, we observe that the EM wave field radiated into the homogeneous, isotropic embedding $\mathcal{D}^{\infty}$, whose EM properties are described by (real-valued and positive) electric permittivity ϵ_0 and magnetic permeability μ_0, has the form of a spherical wave

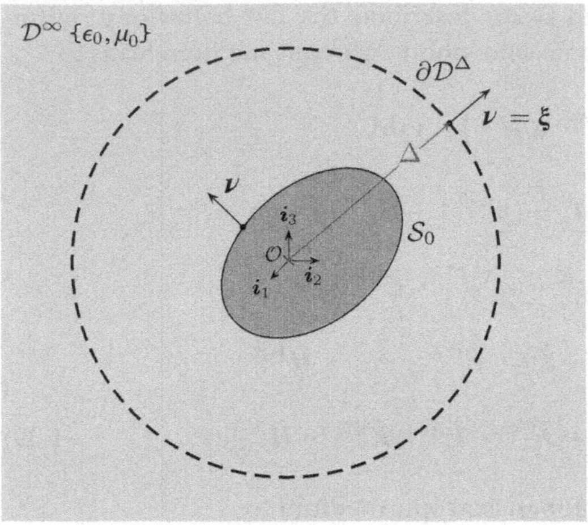

Figure 1.5 Unbounded domain to which the reciprocity theorems are applied.

expanding away from the origin ([16], Sec. 26.11):

$$\{\hat{\boldsymbol{E}}, \hat{\boldsymbol{H}}\}(\boldsymbol{x}, s) = \{\hat{\boldsymbol{E}}^\infty, \hat{\boldsymbol{H}}^\infty\}(\boldsymbol{\xi}, s)\exp(-s|\boldsymbol{x}|/c_0)(4\pi|\boldsymbol{x}|)^{-1} \times \left[1 + O(|\boldsymbol{x}|^{-1})\right] \tag{1.41}$$

as $|\boldsymbol{x}| \to \infty$, where $\{\hat{\boldsymbol{E}}^\infty, \hat{\boldsymbol{H}}^\infty\}$ are the electric- and magnetic-type amplitude radiation characteristics, $\boldsymbol{\xi} = \boldsymbol{x}/|\boldsymbol{x}|$ is the unit vector in the direction of observation, and $c_0 = (\epsilon_0\mu_0)^{-1/2} > 0$ is the EM wave speed. Now, making use of the far-field behavior (1.41) in the surface integral of the time-convolution type, we get

$$\int_{\boldsymbol{x}\in\partial\mathcal{D}^\Delta} \left(\hat{\boldsymbol{E}}^{\mathrm{A}} \times \hat{\boldsymbol{H}}^{\mathrm{B}} - \hat{\boldsymbol{E}}^{\mathrm{B}} \times \hat{\boldsymbol{H}}^{\mathrm{A}}\right) \cdot \boldsymbol{\nu}\,\mathrm{d}A = O(\Delta^{-1}) \tag{1.42}$$

as $\Delta \to \infty$, since the leading terms in the integrand that are of order Δ^{-2} cancel each other. Hence, owing to causality of the interrelated EM wave fields, the time-convolution-type surface integral over $\partial\mathcal{D}^\Delta$ vanishes as $\Delta \to \infty$. On the other hand, the time-correlation type of the surface interaction integral leads to a non-vanishing contribution:

$$\int_{\boldsymbol{x}\in\partial\mathcal{D}^\Delta} \left(\hat{\boldsymbol{E}}^{\mathrm{A}} \times \hat{\boldsymbol{H}}^{\mathrm{B}\circledast} + \hat{\boldsymbol{E}}^{\mathrm{B}\circledast} \times \hat{\boldsymbol{H}}^{\mathrm{A}}\right) \cdot \boldsymbol{\nu}\,\mathrm{d}A = \left(\eta_0/8\pi^2\right) \int_{\boldsymbol{\xi}\in\Omega} \hat{\boldsymbol{E}}^{\mathrm{A};\infty}(\boldsymbol{\xi}, s) \cdot \hat{\boldsymbol{E}}^{\mathrm{B};\infty}(\boldsymbol{\xi}, -s)\mathrm{d}\Omega \times \left[1 + O(\Delta^{-1})\right] \tag{1.43}$$

as $\Delta \to \infty$, where the integration on the right-hand side is carried out over $\Omega = \{\xi \cdot \xi = 1\}$ defining a unit sphere. Consequently, if the medium exterior to $\mathcal{S}_0$ is source-free, we may, in view of its self-adjointness, write

$$\int_{x \in \mathcal{S}_0} \left(\hat{E}^{\mathrm{A}} \times \hat{H}^{\mathrm{B}} - \hat{E}^{\mathrm{B}} \times \hat{H}^{\mathrm{A}} \right) \cdot \nu \, \mathrm{d}A = 0 \tag{1.44}$$

$$\begin{aligned} &\int_{x \in \mathcal{S}_0} \left(\hat{E}^{\mathrm{A}} \times \hat{H}^{\mathrm{B}\circledast} + \hat{E}^{\mathrm{B}\circledast} \times \hat{H}^{\mathrm{A}} \right) \cdot \nu \, \mathrm{d}A \\ &= \left(\eta_0/8\pi^2\right) \int_{\xi \in \Omega} \hat{E}^{\mathrm{A};\infty}(\xi, s) \cdot \hat{E}^{\mathrm{B};\infty}(\xi, -s) \mathrm{d}\Omega \\ &= \left(8\pi^2 \eta_0\right)^{-1} \int_{\xi \in \Omega} \hat{H}^{\mathrm{A};\infty}(\xi, s) \cdot \hat{H}^{\mathrm{B};\infty}(\xi, -s) \mathrm{d}\Omega \end{aligned} \tag{1.45}$$

where we have tacitly taken the limit $\Delta \to \infty$ (see Figure 1.5). In conclusion, the surface integral contribution from the outer bounding surface $\partial\mathcal{D}^{\Delta}$ is vanishing only for the *time-convolution* interaction of two *causal* wave fields. The time-correlation interaction results in the nonzero contribution (1.43) that for the lossless embedding described by two positive scalar constants $\{\epsilon_0, \mu_0\}$ should be approached via the real-frequency domain (see Section 1.3).

1.5 Description of the Antenna Configuration

We will not limit our further analysis to a particular antenna geometry; instead, we will take the advantage of the generic antenna model introduced in Ref. [12], which encompasses all EM antennas used in practice (see Figure 1.6). The antenna system occupies a bounded domain $\mathcal{A} \subset \mathbb{R}^3$ that is terminated by surfaces $\mathcal{S}_0$ and $\mathcal{S}_1$. Surface $\mathcal{S}_0$ separates the antenna system from the exterior domain denoted by $\mathcal{D}^{\infty}$, while surface $\mathcal{S}_1$ represents the terminal surface where the antenna system is accessible via its N-ports. The maximum diameter of the domain enclosed by $\mathcal{S}_1$ is supposed to be small with respect to the pulse time width of the excited EM wave fields. The bounding surfaces $\mathcal{S}_0$ and $\mathcal{S}_1$ may partially overlap.

The antenna system consists of a linear and passive media, whose EM properties can be described by the (tensorial) transverse admittance and the longitudinal impedance, $\underline{\hat{\eta}} = \underline{\hat{\eta}}(x, s)$ and $\underline{\hat{\zeta}} = \underline{\hat{\zeta}}(x, s)$, respectively (see Eqs. (1.23) and (1.24)) These constitutive functions are piecewise continuous functions with respect to the position vector x, that is, they may show finite-jump discontinuities across bounded interfaces, and, in view of the uniqueness theorem given in Chapter 2, they are positive definite tensors of rank 2 for all real and

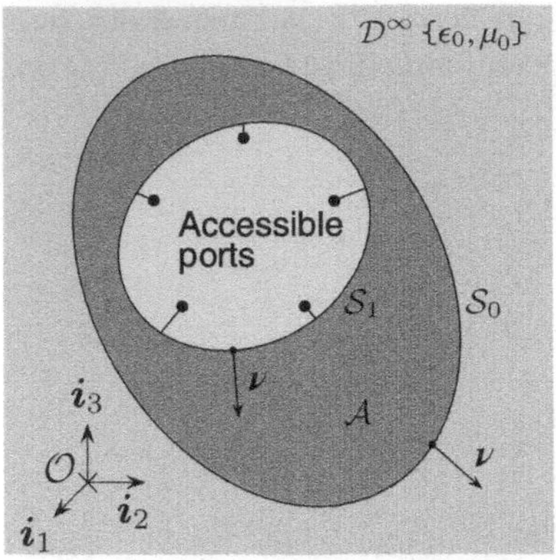

Figure 1.6 Generic antenna configuration.

positive values of s. The antenna system may also contain perfectly conducting surfaces. The antenna structure itself is placed in the linear, homogeneous, and isotropic embedding $\mathcal{D}^\infty$, whose EM properties are defined via its (real-valued and positive) electric permittivity ϵ_0 and magnetic permeability μ_0.

1.5.1 Antenna Power Conservation

The power-reciprocity theorem in antenna theory [18] calls for the application of the reciprocity theorem of the time-correlation type to the domain occupied by the antenna system according to Table 1.3. Taking into account the orientation of the outer normal vector along the bounding surfaces, one may arrive at

Table 1.3 Application of the reciprocity theorem of the time-correlation type

Domain $\mathcal{A}$		
Time-correlation	State (A)	State (A)
Source	0	0
Field	$\{\hat{\boldsymbol{E}}^{\mathrm{A}}, \hat{\boldsymbol{H}}^{\mathrm{A}}\}$	$\{\hat{\boldsymbol{E}}^{\mathrm{A}}, \hat{\boldsymbol{H}}^{\mathrm{A}}\}$
Material	$\{\hat{\underline{\eta}}, \hat{\underline{\zeta}}\}$	$\{\hat{\underline{\eta}}, \hat{\underline{\zeta}}\}$

$$\begin{aligned}
&\int_{x\in\mathcal{S}_0}\left(\hat{E}^{\mathrm{A}}\times\hat{H}^{\mathrm{A}\circledast}+\hat{E}^{\mathrm{A}\circledast}\times\hat{H}^{\mathrm{A}}\right)\cdot\nu\,\mathrm{d}A\\
&=\int_{x\in\mathcal{S}_1}\left(\hat{E}^{\mathrm{A}}\times\hat{H}^{\mathrm{A}\circledast}+\hat{E}^{\mathrm{A}\circledast}\times\hat{H}^{\mathrm{A}}\right)\cdot\nu\,\mathrm{d}A\\
&\quad-\int_{x\in\mathcal{A}}\left\{\hat{H}^{\mathrm{A}}\cdot\left[\underline{\hat{\zeta}}^{\circledast}+\underline{\hat{\zeta}}^{T}\right]\cdot\hat{H}^{\mathrm{A}\circledast}\right.\\
&\qquad\left.+\hat{E}^{\mathrm{A}}\cdot\left[\underline{\hat{\eta}}^{\circledast}+\underline{\hat{\eta}}^{T}\right]\cdot\hat{E}^{\mathrm{A}\circledast}\right\}\mathrm{d}V
\end{aligned}\tag{1.46}$$

In case that antenna's losses can be entirely included in the electric-conduction relaxation function, while its (s-independent) electric permittivity and magnetic permeability relaxation functions are symmetrical tensors of rank 2, Eq. (1.46) has the following form:

$$\begin{aligned}
&\int_{x\in\mathcal{S}_0}\left(\hat{E}^{\mathrm{A}}\times\hat{H}^{\mathrm{A}\circledast}+\hat{E}^{\mathrm{A}\circledast}\times\hat{H}^{\mathrm{A}}\right)\cdot\nu\,\mathrm{d}A\\
&=\int_{x\in\mathcal{S}_1}\left(\hat{E}^{\mathrm{A}}\times\hat{H}^{\mathrm{A}\circledast}+\hat{E}^{\mathrm{A}\circledast}\times\hat{H}^{\mathrm{A}}\right)\cdot\nu\,\mathrm{d}A\\
&\quad-\int_{x\in\mathcal{A}}\left\{\hat{E}^{\mathrm{A}}\cdot\left[\underline{\hat{\sigma}}^{\circledast}+\underline{\hat{\sigma}}^{T}\right]\cdot\hat{E}^{\mathrm{A}\circledast}\right\}\mathrm{d}V
\end{aligned}\tag{1.47}$$

No matter whether the antenna system operates in the transmitting or receiving state, the volume integrals in Eqs. (1.46) and (1.47) are for $\{s=\delta+\mathrm{i}\omega,\delta\downarrow 0,\omega\in\mathbb{R}\}$ proportional to the (time-averaged) power dissipated in between the exterior bounding surface $\mathcal{S}_0$ and the interface surface $\mathcal{S}_1$. In line with Eqs. (1.37) and (1.38), this contribution vanishes whenever the medium in $\mathcal{A}$ is time-reverse self-adjoint in its EM behavior. This includes, in particular, the instantaneously reacting media whose electrical permittivity and magnetic permeability are symmetrical tensors of rank 2, specifically

$$\underline{\epsilon}(x)=\underline{\epsilon}^{T}(x)\tag{1.48}$$

$$\underline{\mu}(x)=\underline{\mu}^{T}(x)\tag{1.49}$$

for all $x\in\mathcal{A}$, as well as (idealized) perfectly electrically conducting (PEC) antenna models. An important class of antennas in this category is represented by a wire antenna. For this antenna system, the external surface $\mathcal{S}_0$ is formed by a closed cylindrical surface closely surrounding the PEC arms of the wire antenna including the (vanishing) volume of the excitation gap. The latter is enclosed by the cylindrical surface $\mathcal{S}_1$ that is at its ends crossed by the antenna ports. Obviously, in this antenna configuration, the terminal surface $\mathcal{S}_1$ partially overlaps with $\mathcal{S}_0$ (see Figure 1.7).

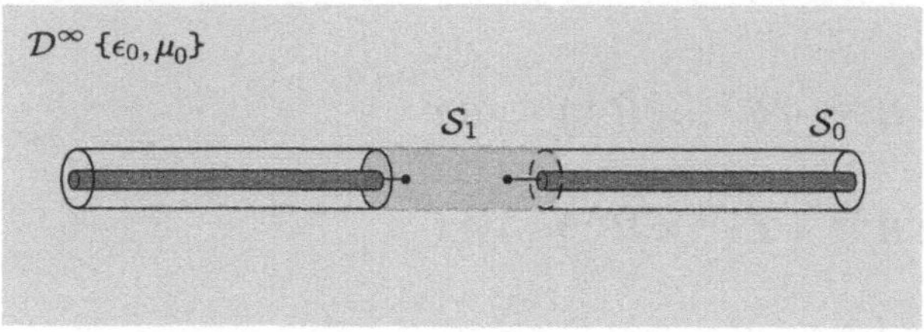

Figure 1.7 Wire antenna and its bounding surfaces.

1.5.2 Antenna Interface Relations

The EM wave quantities on the terminal surface $\mathcal{S}_1$ of the antenna system can be expressed in terms of its Kirchhoff-type quantities (see Ref. [16], Sec. 30.1). To illustrate the procedure that leads to such a relation, we shall associate state (A) from the previous section with the receiving situation (R) in which the antenna system is externally irradiated by the incident EM wave fields. Since the maximum diameter of the domain enclosed by $\mathcal{S}_1$ is small with respect to the pulse time width of the excited fields, the electric field strength $\hat{\boldsymbol{E}}^{\mathrm{R}}$ can be expressed as (the opposite of) the gradient of the scalar potential $\hat{\phi}^{\mathrm{R}}$. Consequently, the surface integral over $\mathcal{S}_1$ from Eq. (1.46) can be written as

$$
\begin{aligned}
&\int_{\boldsymbol{x}\in\mathcal{S}_1} \left(\hat{\boldsymbol{E}}^{\mathrm{R}} \times \hat{\boldsymbol{H}}^{\mathrm{R}\circledast} + \hat{\boldsymbol{E}}^{\mathrm{R}\circledast} \times \hat{\boldsymbol{H}}^{\mathrm{R}}\right) \cdot \boldsymbol{\nu}\,\mathrm{d}A \\
&\simeq -\int_{\boldsymbol{x}\in\mathcal{S}_1} \left(\nabla\hat{\phi}^{\mathrm{R}} \times \hat{\boldsymbol{H}}^{\mathrm{R}\circledast} + \nabla\hat{\phi}^{\mathrm{R}\circledast} \times \hat{\boldsymbol{H}}^{\mathrm{R}}\right) \cdot \boldsymbol{\nu}\,\mathrm{d}A \\
&= -\int_{\boldsymbol{x}\in\mathcal{S}_1} \left[\nabla\times(\hat{\phi}^{\mathrm{R}}\hat{\boldsymbol{H}}^{\mathrm{R}\circledast}) + \nabla\times(\hat{\phi}^{\mathrm{R}\circledast}\hat{\boldsymbol{H}}^{\mathrm{R}})\right] \cdot \boldsymbol{\nu}\,\mathrm{d}A \\
&\quad + \int_{\boldsymbol{x}\in\mathcal{S}_1} \left[\hat{\phi}^{\mathrm{R}}(\nabla\times\hat{\boldsymbol{H}}^{\mathrm{R}\circledast}) + \hat{\phi}^{\mathrm{R}\circledast}(\nabla\times\hat{\boldsymbol{H}}^{\mathrm{R}})\right] \cdot \boldsymbol{\nu}\,\mathrm{d}A
\end{aligned}
\tag{1.50}
$$

where we have used integration by parts and $\simeq$ indicates here the low-frequency approximation. Since the first integral on the right-hand side is in view of Stokes' theorem zero, we may, upon using the first Maxwell's equation (see Eq. (1.21)) in the second one, write

$$
\begin{aligned}
&\int_{\boldsymbol{x}\in\mathcal{S}_1} \left(\hat{\boldsymbol{E}}^{\mathrm{R}} \times \hat{\boldsymbol{H}}^{\mathrm{R}\circledast} + \hat{\boldsymbol{E}}^{\mathrm{R}\circledast} \times \hat{\boldsymbol{H}}^{\mathrm{R}}\right) \cdot \boldsymbol{\nu}\,\mathrm{d}A \\
&\qquad \simeq \int_{\boldsymbol{x}\in\mathcal{S}_1} \left(\hat{\phi}^{\mathrm{R}}\,\boldsymbol{\nu}\cdot\hat{\boldsymbol{J}}^{\mathrm{R}\circledast} + \hat{\phi}^{\mathrm{R}\circledast}\,\boldsymbol{\nu}\cdot\hat{\boldsymbol{J}}^{\mathrm{R}}\right)\mathrm{d}A \\
&\qquad\qquad \simeq -\sum_{n=1}^{N}\left[\hat{V}_n^{\mathrm{R}}(s)\hat{I}_n^{\mathrm{R}}(-s) + \hat{V}_n^{\mathrm{R}}(-s)\hat{I}_n^{\mathrm{R}}(s)\right]
\end{aligned}
\tag{1.51}
$$

where we have used the fact that the electric current volume density on S_1 is dominated by the conduction current flowing in the perfect conductors of the N-port termination and that these electric currents are in the receiving state oriented into the load. Furthermore, in the interior of the terminal surface S_1, we have chosen a reference point where the scalar potential $\hat{\phi}^{\mathrm{R}}$ has the value zero. Consequently, the scalar potential at the nth PEC port, that is related to this reference point, is denoted as $\hat{V}_n^{\mathrm{R}}$. Employing these results, the time-averaged power absorbed by the antenna load can be written as

$$\hat{P}^{\mathrm{L}}(s) = \tfrac{1}{4}\sum_{n=1}^{N}\left[\hat{V}_n^{\mathrm{R}}(s)\hat{I}_n^{\mathrm{R}}(-s) + \hat{V}_n^{\mathrm{R}}(-s)\hat{I}_n^{\mathrm{R}}(s)\right]$$
$$\simeq -\tfrac{1}{4}\int_{\boldsymbol{x}\in S_1}\left(\hat{\boldsymbol{E}}^{\mathrm{R}} \times \hat{\boldsymbol{H}}^{\mathrm{R}\circledast} + \hat{\boldsymbol{E}}^{\mathrm{R}\circledast} \times \hat{\boldsymbol{H}}^{\mathrm{R}}\right)\cdot \boldsymbol{\nu}\,\mathrm{d}A \tag{1.52}$$

for $\{s = \delta + \mathrm{i}\omega, \delta \downarrow 0, \omega \in \mathbb{R}\}$. Equation (1.52) thus makes possible to clearly interpret the integral over S_1 in the antenna power conservation relations (1.46) and (1.47) in the receiving (R) state.

2

Antenna Uniqueness Theorem

The main purpose of this chapter is to determine what conditions lead, for a given antenna excitation, to only one and the same (causal) radiated EM wave field. The standard procedure to solve the problem is based on the time-domain Poynting's theorem ([39], Sec. 9.2). For the general type of dispersive media constituting the analyzed antenna system (see Section 1.5), however, this way is not generally applicable. An elegant remedy of this situation has been suggested by de Hoop [17], whose proof relies on Lerch's uniqueness theorem for the one-sided Laplace transformation with the real-valued and positive parameter s. Following the lines of reasoning introduced in Ref. [17], it will be next demonstrated that the transmitting antenna problem has the unique solution provided that the antenna constitutive parameters are positive definite tensors for all real and positive values of s.

2.1 Problem Description

We shall analyze the N-port antenna system that is excited by electric current pulses $I_n^{\mathrm{T}} = I_n^{\mathrm{T}}(t)$ at its accessible ports (see Figure 2.1). Namely, under the assumption that one and the same antenna excitation, $I_n^{\mathrm{T}} = I_n^{\tilde{\mathrm{T}}}$ for all $n \in \{1, \ldots, N\}$, activates two distinct EM wave fields $\{\boldsymbol{E}^{\mathrm{T}}, \boldsymbol{H}^{\mathrm{T}}\}$ and $\{\boldsymbol{E}^{\tilde{\mathrm{T}}}, \boldsymbol{H}^{\tilde{\mathrm{T}}}\}$, we will determine the conditions that lead to the contradiction.

2.2 Problem Solution

The EM wave fields radiated by the transmitting antenna system in states (T) and ($\tilde{\mathrm{T}}$) are governed by the EM field equations (cf. Eqs. (1.21) and (1.22))

$$\nabla \times \hat{\boldsymbol{H}}^{\mathrm{T},\tilde{\mathrm{T}}} - \underline{\hat{\eta}} \cdot \hat{\boldsymbol{E}}^{\mathrm{T},\tilde{\mathrm{T}}} = \boldsymbol{0} \tag{2.1}$$

$$\nabla \times \hat{\boldsymbol{E}}^{\mathrm{T},\tilde{\mathrm{T}}} + \underline{\hat{\zeta}} \cdot \hat{\boldsymbol{H}}^{\mathrm{T},\tilde{\mathrm{T}}} = \boldsymbol{0} \tag{2.2}$$

Electromagnetic Reciprocity in Antenna Theory, First Edition. Martin Stumpf.

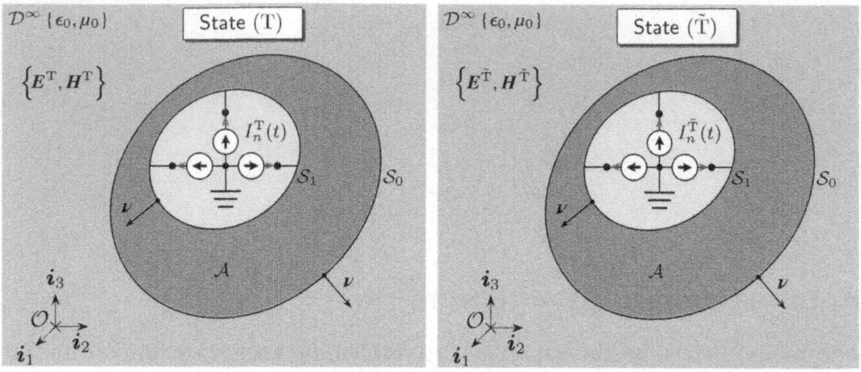

Figure 2.1 Transmitting situations to which the uniqueness theorem applies. Here we take $I_n^{\mathrm{T}} = I_n^{\tilde{\mathrm{T}}}$ for all $n \in \{1, 2, 3\}$.

for $\boldsymbol{x} \in \mathcal{A}$, where $\underline{\hat{\eta}}$ and $\underline{\hat{\zeta}}$ are the transverse admittance and the longitudinal impedance (per length of the medium) describing the EM properties of the medium in the bounded domain occupied by the antenna system (see Eqs. (1.23) and (1.24)). The antenna system then radiates into the embedding where the transmitted EM fields satisfy the source-free EM wave equations:

$$\nabla \times \hat{\boldsymbol{H}}^{\mathrm{T},\tilde{\mathrm{T}}} - s\epsilon_0 \hat{\boldsymbol{E}}^{\mathrm{T},\tilde{\mathrm{T}}} = \boldsymbol{0} \tag{2.3}$$

$$\nabla \times \hat{\boldsymbol{E}}^{\mathrm{T},\tilde{\mathrm{T}}} + s\mu_0 \hat{\boldsymbol{H}}^{\mathrm{T},\tilde{\mathrm{T}}} = \boldsymbol{0} \tag{2.4}$$

for $\boldsymbol{x} \in \mathcal{D}^\infty$. In virtue of causality, the radiated EM wave field quantities are bounded as $|\boldsymbol{x}| \to \infty$. The decisive point in a proof of uniqueness is the demonstration that the corresponding homogeneous problem does not possess a nonidentically vanishing solution. To this end, we define the field difference:

$$\{\Delta\hat{\boldsymbol{E}}^{\mathrm{T}}, \Delta\hat{\boldsymbol{H}}^{\mathrm{T}}\}(\boldsymbol{x}, s) \triangleq \{\hat{\boldsymbol{E}}^{\tilde{\mathrm{T}}} - \hat{\boldsymbol{E}}^{\mathrm{T}}, \hat{\boldsymbol{H}}^{\tilde{\mathrm{T}}} - \hat{\boldsymbol{H}}^{\mathrm{T}}\}(\boldsymbol{x}, s) \tag{2.5}$$

for $\boldsymbol{x} \in \mathbb{R}^3$ that obviously satisfies the same EM field equations as the radiated field (namely, Eqs. (2.1)–(2.4)) in both states. Next, combining the field equations in the antenna domain, one finds

$$\int_{\boldsymbol{x}\in\mathcal{S}_0} \left(\Delta\hat{\boldsymbol{E}}^{\mathrm{T}} \times \Delta\hat{\boldsymbol{H}}^{\mathrm{T}}\right) \cdot \boldsymbol{\nu}\,\mathrm{d}A = \int_{\boldsymbol{x}\in\mathcal{S}_1} \left(\Delta\hat{\boldsymbol{E}}^{\mathrm{T}} \times \Delta\hat{\boldsymbol{H}}^{\mathrm{T}}\right) \cdot \boldsymbol{\nu}\,\mathrm{d}A$$
$$- \int_{\boldsymbol{x}\in\mathcal{A}} \left(\Delta\hat{\boldsymbol{E}}^{\mathrm{T}} \cdot \underline{\hat{\eta}} \cdot \Delta\hat{\boldsymbol{E}}^{\mathrm{T}} + \Delta\hat{\boldsymbol{H}}^{\mathrm{T}} \cdot \underline{\hat{\zeta}} \cdot \Delta\hat{\boldsymbol{H}}^{\mathrm{T}}\right) \mathrm{d}V \tag{2.6}$$

where the surface integral along the antenna interfacing surface $\mathcal{S}_1$ is, thanks to the one and the same excitation applied in both transmitting situations, identically zero, that is,

$$\int_{x \in S_1} \left(\Delta \hat{E}^{\mathrm{T}} \times \Delta \hat{H}^{\mathrm{T}} \right) \cdot \nu \, \mathrm{d}A \simeq \sum_{n=1}^{N} \Delta \hat{V}_n^{\mathrm{T}}(s) \Delta \hat{I}_n^{\mathrm{T}}(s) = 0 \tag{2.7}$$

where $\Delta \hat{I}_n^{\mathrm{T}} = \hat{I}_n^{\tilde{\mathrm{T}}} - \hat{I}_n^{\mathrm{T}}$ and $\Delta \hat{V}_n^{\mathrm{T}} = \hat{V}_n^{\tilde{\mathrm{T}}} - \hat{V}_n^{\mathrm{T}}$. The same procedure applied to the unbounded domain exterior to the antenna system yields

$$\begin{aligned} &\int_{x \in S_0} \left(\Delta \hat{E}^{\mathrm{T}} \times \Delta \hat{H}^{\mathrm{T}} \right) \cdot \nu \, \mathrm{d}A \\ &= s \int_{x \in D^{\infty}} \left(\epsilon_0 \Delta \hat{E}^{\mathrm{T}} \cdot \Delta \hat{E}^{\mathrm{T}} + \mu_0 \Delta \hat{H}^{\mathrm{T}} \cdot \Delta \hat{H}^{\mathrm{T}} \right) \mathrm{d}V \end{aligned} \tag{2.8}$$

where we have used the limiting process described in Section 1.4.3. Upon combining Eqs. (2.6)–(2.8), we get

$$\begin{aligned} &s \int_{x \in D^{\infty}} \left(\epsilon_0 \Delta \hat{E}^{\mathrm{T}} \cdot \Delta \hat{E}^{\mathrm{T}} + \mu_0 \Delta \hat{H}^{\mathrm{T}} \cdot \Delta \hat{H}^{\mathrm{T}} \right) \mathrm{d}V \\ &= - \int_{x \in \mathcal{A}} \left(\Delta \hat{E}^{\mathrm{T}} \cdot \underline{\hat{\eta}} \cdot \Delta \hat{E}^{\mathrm{T}} + \Delta \hat{H}^{\mathrm{T}} \cdot \underline{\hat{\zeta}} \cdot \Delta \hat{H}^{\mathrm{T}} \right) \mathrm{d}V \end{aligned} \tag{2.9}$$

Now it is observed that the left-hand side of Eq. (2.9) is along Lerch's sequence $\mathcal{L} = \{s \in \mathbb{R}; s = s_0 + nh, h > 0, n = 1, 2, \ldots\}$ real-valued and positive for any non identically vanishing $\Delta \hat{E}^{\mathrm{T}}$ and $\Delta \hat{H}^{\mathrm{T}}$. As long as $\underline{\hat{\eta}}$ and $\underline{\hat{\zeta}}$ are positive definite tensors of rank 2 for all $x \in \mathcal{A}$ and $s \in \mathcal{L}$, Eq. (2.9) leads to a contradiction unless $\Delta \hat{E}^{\mathrm{T}} = 0$ and $\Delta \hat{H}^{\mathrm{T}} = 0$ for all $x \in \mathcal{A} \cup D^{\infty}$. The latter thus implies the unique solution:

$$\left\{ \hat{E}^{\mathrm{T}}, \hat{H}^{\mathrm{T}} \right\}(x, s) = \left\{ \hat{E}^{\tilde{\mathrm{T}}}, \hat{H}^{\tilde{\mathrm{T}}} \right\}(x, s) \tag{2.10}$$

throughout the problem configuration, that is, for all $x \in \mathcal{A} \cup D^{\infty}$ and $s \in \mathcal{L}$. The Lerch uniqueness theorem for the one-sided Laplace transformation then offers the possibility of rewriting Eq. (2.10) in terms of the corresponding *causal* wave fields

$$\left\{ E^{\mathrm{T}}, H^{\mathrm{T}} \right\}(x, t) = \left\{ E^{\tilde{\mathrm{T}}}, H^{\tilde{\mathrm{T}}} \right\}(x, t) \tag{2.11}$$

for all $x \in \mathcal{A} \cup D^{\infty}$ and $t > 0$. In conclusion, it has been shown that there is only one and the same (causal) radiated EM wave field in the problem configuration provided that the constitutive relaxation functions describing the EM behavior of the antenna system are positive definite tensors of rank 2 along the Lerch sequence (and hence along the real positive axis $\{s \in \mathbb{R}; s > s_0\}$).

3

Forward-Scattering Theorem in Antenna Theory

3.1 Problem Description

In this chapter, we shall discuss the forward-scattering theorem concerning the N-port receiving antenna system shown in Figure 3.1. The antenna system is placed in the linear, homogeneous, and isotropic embedding whose EM properties are described by real-valued and positive constants ϵ_0 and μ_0.

The antenna system under consideration is irradiated by the uniform EM plane wave, namely,

$$\hat{\boldsymbol{E}}^{\mathrm{i}}(\boldsymbol{x}, s) = \boldsymbol{\alpha}\, \hat{e}^{\mathrm{i}}(s) \exp(-s\boldsymbol{\beta} \cdot \boldsymbol{x}/c_0) \tag{3.1}$$

$$\hat{\boldsymbol{H}}^{\mathrm{i}}(\boldsymbol{x}, s) = (\boldsymbol{\beta} \times \boldsymbol{\alpha})\, \eta_0 \hat{e}^{\mathrm{i}}(s) \exp(-s\boldsymbol{\beta} \cdot \boldsymbol{x}/c_0) \tag{3.2}$$

where $\boldsymbol{\alpha}$ is a unit vector in the direction of polarization, $\boldsymbol{\beta}$ is a unit vector in the direction of propagation, $c_0 = (\epsilon_0\mu_0)^{-1/2}$ is the EM wave speed, and $\eta_0 = (\epsilon_0/\mu_0)^{1/2}$ is the EM wave admittance.

3.2 Problem Solution

The presence of the antenna system is accounted for via the scattered EM wave field that is defined as the difference between the total EM field in the configuration, $\{\hat{\boldsymbol{E}}^{\mathrm{R}}, \hat{\boldsymbol{H}}^{\mathrm{R}}\}$, and the incident field, that is,

$$\{\hat{\boldsymbol{E}}^{\mathrm{s}}, \hat{\boldsymbol{H}}^{\mathrm{s}}\}(\boldsymbol{x}, s) \triangleq \{\hat{\boldsymbol{E}}^{\mathrm{R}}, \hat{\boldsymbol{H}}^{\mathrm{R}}\}(\boldsymbol{x}, s) - \{\hat{\boldsymbol{E}}^{\mathrm{i}}, \hat{\boldsymbol{H}}^{\mathrm{i}}\}(\boldsymbol{x}, s) \tag{3.3}$$

Accordingly, the incident field can be interpreted as the EM field that would exist in the problem configuration if there were no antenna system. The scattered field admits the far-field representation for causal EM wave fields ([16], Sec. 26.11):

$$\{\hat{\boldsymbol{E}}^{\mathrm{s}}, \hat{\boldsymbol{H}}^{\mathrm{s}}\}(\boldsymbol{x}, s) = \{\hat{\boldsymbol{E}}^{\mathrm{s};\infty}, \hat{\boldsymbol{H}}^{\mathrm{s};\infty}\}(\boldsymbol{\xi}, s) \exp(-s|\boldsymbol{x}|/c_0)(4\pi|\boldsymbol{x}|)^{-1} \times \left[1 + O(|\boldsymbol{x}|^{-1})\right] \tag{3.4}$$

Electromagnetic Reciprocity in Antenna Theory, First Edition. Martin Stumpf.

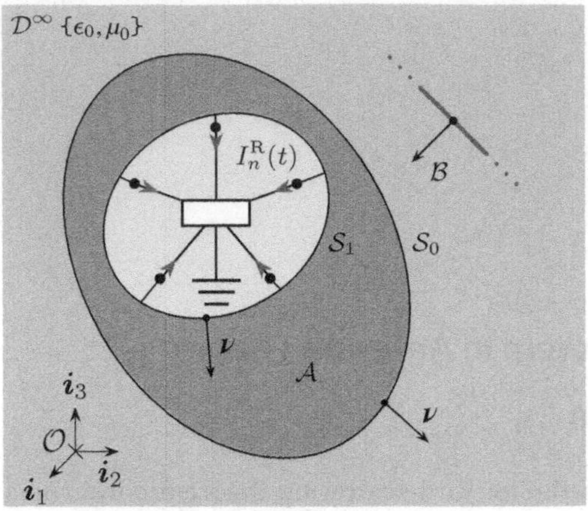

Figure 3.1 Receiving antenna system.

as $|\boldsymbol{x}| \to \infty$, which allows–using Huygens' surface integral representation of the scattered field–expressing its far-field amplitude in terms of the equivalent electric and magnetic current surface densities on the bounding surface of the antenna system (cf. Ref. [11], Eq. (3.8)):

$$\hat{\boldsymbol{E}}^{s;\infty}(\boldsymbol{\xi}, s) = s\mu_0(\boldsymbol{\xi}\boldsymbol{\xi} - \underline{\boldsymbol{I}}) \int_{\boldsymbol{x}\in\mathcal{S}_0} \exp(s\boldsymbol{\xi} \cdot \boldsymbol{x}/c_0)\big[\boldsymbol{\nu}(\boldsymbol{x}) \times \hat{\boldsymbol{H}}^s(\boldsymbol{x}, s)\big]\mathrm{d}A$$
$$+ sc_0^{-1}\boldsymbol{\xi} \times \int_{\boldsymbol{x}\in\mathcal{S}_0} \exp(s\boldsymbol{\xi} \cdot \boldsymbol{x}/c_0)\big[\hat{\boldsymbol{E}}^s(\boldsymbol{x}, s) \times \boldsymbol{\nu}(\boldsymbol{x})\big]\mathrm{d}A \qquad (3.5)$$

for all $\boldsymbol{\xi} \in \Omega = \{\boldsymbol{\xi} \cdot \boldsymbol{\xi} = 1\}$ on the unit sphere. Consequently, combination of Eqs. (3.1) and (3.2) with (3.5) yields

$$\int_{\boldsymbol{x}\in\mathcal{S}_0} \left(\hat{\boldsymbol{E}}^{i\circledast} \times \hat{\boldsymbol{H}}^s + \hat{\boldsymbol{E}}^s \times \hat{\boldsymbol{H}}^{i\circledast}\right) \cdot \boldsymbol{\nu}\,\mathrm{d}A$$
$$= \hat{e}^i(-s)\,\boldsymbol{\alpha} \cdot \hat{\boldsymbol{E}}^{s;\infty}(\boldsymbol{\beta}, s)/s\mu_0 \qquad (3.6)$$

$$\int_{\boldsymbol{x}\in\mathcal{S}_0} \left(\hat{\boldsymbol{E}}^{i} \times \hat{\boldsymbol{H}}^{s\circledast} + \hat{\boldsymbol{E}}^{s\circledast} \times \hat{\boldsymbol{H}}^{i}\right) \cdot \boldsymbol{\nu}\,\mathrm{d}A$$
$$= -\hat{e}^i(s)\,\boldsymbol{\alpha} \cdot \hat{\boldsymbol{E}}^{s;\infty}(\boldsymbol{\beta}, -s)/s\mu_0 \qquad (3.7)$$

Making use of Eq. (3.3), it is observed that

$$\begin{aligned}&\hat{\boldsymbol{E}}^{\mathrm{i}\circledast} \times \hat{\boldsymbol{H}}^{\mathrm{s}} + \hat{\boldsymbol{E}}^{\mathrm{s}} \times \hat{\boldsymbol{H}}^{\mathrm{i}\circledast} + \hat{\boldsymbol{E}}^{\mathrm{i}} \times \hat{\boldsymbol{H}}^{\mathrm{s}\circledast} + \hat{\boldsymbol{E}}^{\mathrm{s}\circledast} \times \hat{\boldsymbol{H}}^{\mathrm{i}}\\ &= \hat{\boldsymbol{E}}^{\mathrm{R}} \times \hat{\boldsymbol{H}}^{\mathrm{R}\circledast} + \hat{\boldsymbol{E}}^{\mathrm{R}\circledast} \times \hat{\boldsymbol{H}}^{\mathrm{R}} - (\hat{\boldsymbol{E}}^{\mathrm{s}} \times \hat{\boldsymbol{H}}^{\mathrm{s}\circledast} + \hat{\boldsymbol{E}}^{\mathrm{s}\circledast} \times \hat{\boldsymbol{H}}^{\mathrm{s}})\\ &- (\hat{\boldsymbol{E}}^{\mathrm{i}} \times \hat{\boldsymbol{H}}^{\mathrm{i}\circledast} + \hat{\boldsymbol{E}}^{\mathrm{i}\circledast} \times \hat{\boldsymbol{H}}^{\mathrm{i}})\end{aligned} \tag{3.8}$$

which motivates to define

$$\hat{P}^{\mathrm{a}}(s) \triangleq -\tfrac{1}{2} \int_{\boldsymbol{x}\in S_0} \left(\hat{\boldsymbol{E}}^{\mathrm{R}} \times \hat{\boldsymbol{H}}^{\mathrm{R}\circledast} + \hat{\boldsymbol{E}}^{\mathrm{R}\circledast} \times \hat{\boldsymbol{H}}^{\mathrm{R}}\right) \cdot \boldsymbol{\nu}\,\mathrm{d}A \tag{3.9}$$

$$\hat{P}^{\mathrm{s}}(s) \triangleq \tfrac{1}{2} \int_{\boldsymbol{x}\in S_0} \left(\hat{\boldsymbol{E}}^{\mathrm{s}} \times \hat{\boldsymbol{H}}^{\mathrm{s}\circledast} + \hat{\boldsymbol{E}}^{\mathrm{s}\circledast} \times \hat{\boldsymbol{H}}^{\mathrm{s}}\right) \cdot \boldsymbol{\nu}\,\mathrm{d}A \tag{3.10}$$

being the quantities that in the limit $s = \delta + \mathrm{i}\omega$, $\delta \downarrow 0$, are proportional[1] to the (time-averaged) total power absorbed and scattered by the antenna system, respectively. Also, thanks to the time-reverse self-adjointness of the embedding, we have

$$\int_{\boldsymbol{x}\in S_0} \left(\hat{\boldsymbol{E}}^{\mathrm{i}} \times \hat{\boldsymbol{H}}^{\mathrm{i}\circledast} + \hat{\boldsymbol{E}}^{\mathrm{i}\circledast} \times \hat{\boldsymbol{H}}^{\mathrm{i}}\right) \cdot \boldsymbol{\nu}\,\mathrm{d}A = 0 \tag{3.11}$$

Hence, upon combining Eqs. (3.6)–(3.11), we arrive at the desired relation:

$$\begin{aligned}&\hat{P}^{\mathrm{a}}(s) + \hat{P}^{\mathrm{s}}(s)\\ &= \tfrac{1}{2}\left[\hat{e}^{\mathrm{i}}(s)\,\boldsymbol{\alpha} \cdot \hat{\boldsymbol{E}}^{\mathrm{s};\infty}(\boldsymbol{\beta}, -s) - \hat{e}^{\mathrm{i}}(-s)\,\boldsymbol{\alpha} \cdot \hat{\boldsymbol{E}}^{\mathrm{s};\infty}(\boldsymbol{\beta}, s)\right]/s\mu_0\end{aligned} \tag{3.12}$$

which is the general form of the forward-scattering theorem in the complex-frequency domain. The theorem relates the sum of the scattered and absorbed power by the receiving antenna system to the scattered far-field amplitude observed in the direction of propagation of the incident plane wave (i.e., in the forward direction).

The terms on the left-hand side of Eq. (3.12) can be further elaborated (see also Exercises in Chapter 9). To this end, the reciprocity theorem of the time-correlation type is first applied to the domain enclosed by S_0 and S_1 and to the total field states (see Table 3.1). Making use of Eq. (1.36), this yields

---

1 The equality to the time-averaged power of time-harmonic fields would require to change the factor $\frac{1}{2}$ to $\frac{1}{4}$.

Table 3.1 Application of the reciprocity theorem.

Domain $\mathcal{A}$		
Time-correlation	State (R)	State (R)
Source	0	0
Field	$\{\hat{\boldsymbol{E}}^{\mathrm{R}}, \hat{\boldsymbol{H}}^{\mathrm{R}}\}$	$\{\hat{\boldsymbol{E}}^{\mathrm{R}}, \hat{\boldsymbol{H}}^{\mathrm{R}}\}$
Material	$\{\underline{\hat{\eta}}, \underline{\hat{\zeta}}\}$	$\{\underline{\hat{\eta}}, \underline{\hat{\zeta}}\}$

$$\begin{aligned}
&\int_{\boldsymbol{x}\in\mathcal{S}_0} \left(\hat{\boldsymbol{E}}^{\mathrm{R}} \times \hat{\boldsymbol{H}}^{\mathrm{R}\circledast} + \hat{\boldsymbol{E}}^{\mathrm{R}\circledast} \times \hat{\boldsymbol{H}}^{\mathrm{R}}\right) \cdot \boldsymbol{\nu}\,\mathrm{d}A \\
&= \int_{\boldsymbol{x}\in\mathcal{S}_1} \left(\hat{\boldsymbol{E}}^{\mathrm{R}} \times \hat{\boldsymbol{H}}^{\mathrm{R}\circledast} + \hat{\boldsymbol{E}}^{\mathrm{R}\circledast} \times \hat{\boldsymbol{H}}^{\mathrm{R}}\right) \cdot \boldsymbol{\nu}\,\mathrm{d}A \\
&+ s \int_{\boldsymbol{x}\in\mathcal{A}} \left[\hat{\boldsymbol{H}}^{\mathrm{R}} \cdot \left(\underline{\hat{\mu}}^{\circledast} - \underline{\hat{\mu}}^{\mathcal{T}}\right) \cdot \hat{\boldsymbol{H}}^{\mathrm{R}\circledast} + \hat{\boldsymbol{E}}^{\mathrm{R}} \cdot \left(\underline{\hat{\epsilon}}^{\circledast} - \underline{\hat{\epsilon}}^{\mathcal{T}}\right) \cdot \hat{\boldsymbol{E}}^{\mathrm{R}\circledast}\right] \mathrm{d}V \\
&- \int_{\boldsymbol{x}\in\mathcal{A}} \hat{\boldsymbol{E}}^{\mathrm{R}} \cdot \left(\underline{\hat{\sigma}}^{\circledast} + \underline{\hat{\sigma}}^{\mathcal{T}}\right) \cdot \hat{\boldsymbol{E}}^{\mathrm{R}\circledast} \mathrm{d}V
\end{aligned} \tag{3.13}$$

while the interfacing relation follows (cf. Eq. (1.50)):

$$\begin{aligned}
&\int_{\boldsymbol{x}\in\mathcal{S}_1} \left(\hat{\boldsymbol{E}}^{\mathrm{R}} \times \hat{\boldsymbol{H}}^{\mathrm{R}\circledast} + \hat{\boldsymbol{E}}^{\mathrm{R}\circledast} \times \hat{\boldsymbol{H}}^{\mathrm{R}}\right) \cdot \boldsymbol{\nu}\,\mathrm{d}A \\
&\qquad \simeq -\sum_{n=1}^{N} \left[\hat{V}_n^{\mathrm{R}}(s)\hat{I}_n^{\mathrm{R}}(-s) + \hat{V}_n^{\mathrm{R}}(-s)\hat{I}_n^{\mathrm{R}}(s)\right]
\end{aligned} \tag{3.14}$$

Next, a large group of antenna systems can be described by constitutive parameters that are symmetrical such that $\underline{\hat{\epsilon}}^{\circledast} = \underline{\hat{\epsilon}}^{\mathcal{T}}$ and $\underline{\hat{\mu}}^{\circledast} = \underline{\hat{\mu}}^{\mathcal{T}}$. Accordingly, Eq. (3.9) has the following form

$$\begin{aligned}
\hat{P}^{\mathrm{a}}(s) &= \tfrac{1}{2} \int_{\boldsymbol{x}\in\mathcal{A}} \hat{\boldsymbol{E}}^{\mathrm{R}} \cdot \left(\underline{\hat{\sigma}}^{\circledast} + \underline{\hat{\sigma}}^{\mathcal{T}}\right) \cdot \hat{\boldsymbol{E}}^{\mathrm{R}\circledast} \mathrm{d}V \\
&+ \tfrac{1}{2} \sum_{n=1}^{N} \left[\hat{V}_n^{\mathrm{R}}(s)\hat{I}_n^{\mathrm{R}}(-s) + \hat{V}_n^{\mathrm{R}}(-s)\hat{I}_n^{\mathrm{R}}(s)\right]
\end{aligned} \tag{3.15}$$

that can be related to the power dissipated in the antenna body and in the antenna load. Furthermore, the reciprocity theorem of the time-correlation type is applied to the unbounded domain exterior to the antenna system and to the scattered field state (see Table 3.2). Employing Eq. (1.45) that applies to the

Table 3.2 Application of the reciprocity theorem.

Domain $\mathcal{D}^\infty$		
Time-correlation	State (s)	State (s)
Source	0	0
Field	$\{\hat{\boldsymbol{E}}^{\mathrm{s}}, \hat{\boldsymbol{H}}^{\mathrm{s}}\}$	$\{\hat{\boldsymbol{E}}^{\mathrm{s}}, \hat{\boldsymbol{H}}^{\mathrm{s}}\}$
Material	$\{\epsilon_0, \mu_0\}$	$\{\epsilon_0, \mu_0\}$

linear, homogeneous, and isotropic embedding and to causal EM wave fields, we arrive at

$$\int_{\boldsymbol{x}\in\mathcal{S}_0} \left(\hat{\boldsymbol{E}}^{\mathrm{s}} \times \hat{\boldsymbol{H}}^{\mathrm{s}\circledast} + \hat{\boldsymbol{E}}^{\mathrm{s}\circledast} \times \hat{\boldsymbol{H}}^{\mathrm{s}}\right) \cdot \boldsymbol{\nu}\,\mathrm{d}A$$
$$= \left(\eta_0/8\pi^2\right) \int_{\boldsymbol{\xi}\in\Omega} \hat{\boldsymbol{E}}^{\mathrm{s};\infty}(\boldsymbol{\xi}, s) \cdot \hat{\boldsymbol{E}}^{\mathrm{s};\infty}(\boldsymbol{\xi}, -s)\mathrm{d}\Omega \tag{3.16}$$

that makes possible to rewrite Eq. (3.10) as

$$\hat{P}^{\mathrm{s}}(s) = \left(\eta_0/16\pi^2\right) \int_{\boldsymbol{\xi}\in\Omega} \hat{\boldsymbol{E}}^{\mathrm{s};\infty}(\boldsymbol{\xi}, s) \cdot \hat{\boldsymbol{E}}^{\mathrm{s};\infty}(\boldsymbol{\xi}, -s)\mathrm{d}\Omega \tag{3.17}$$

The latter expressions can be substituted in the general form of the forward-scattering theorem (3.12) to get its special form applying to the N-port antenna system (cf. Ref. [12], Eq. (39)):

$$\frac{1}{2} \int_{\boldsymbol{x}\in\mathcal{A}} \hat{\boldsymbol{E}}^{\mathrm{R}} \cdot \left(\underline{\hat{\boldsymbol{\sigma}}}^{\circledast} + \underline{\hat{\boldsymbol{\sigma}}}^{T}\right) \cdot \hat{\boldsymbol{E}}^{\mathrm{R}\circledast}\mathrm{d}V$$
$$+ \frac{1}{2} \sum_{n=1}^{N} \left[\hat{V}_n^{\mathrm{R}}(s)\hat{I}_n^{\mathrm{R}}(-s) + \hat{V}_n^{\mathrm{R}}(-s)\hat{I}_n^{\mathrm{R}}(s)\right]$$
$$+ \left(\eta_0/16\pi^2\right) \int_{\boldsymbol{\xi}\in\Omega} \hat{\boldsymbol{E}}^{\mathrm{s};\infty}(\boldsymbol{\xi}, s) \cdot \hat{\boldsymbol{E}}^{\mathrm{s};\infty}(\boldsymbol{\xi}, -s)\mathrm{d}\Omega$$
$$= \frac{1}{2}\left[\hat{e}^{\mathrm{i}}(s)\,\boldsymbol{\alpha} \cdot \hat{\boldsymbol{E}}^{\mathrm{s};\infty}(\boldsymbol{\beta}, -s) - \hat{e}^{\mathrm{i}}(-s)\,\boldsymbol{\alpha} \cdot \hat{\boldsymbol{E}}^{\mathrm{s};\infty}(\boldsymbol{\beta}, s)\right]/s\mu_0 \tag{3.18}$$

Exercises

- Express the general form of the forward-scattering theorem (3.12) in the real-frequency domain.

Hint: Multiply Eq. (3.12) by factor $\frac{1}{2}$ and take the limit $s = \delta + \mathrm{i}\omega$, $\delta \downarrow 0$. Consequently, interpret the right-hand sides of

$$\tfrac{1}{2}\hat{P}^{\mathrm{a}}(\mathrm{i}\omega) = -\tfrac{1}{2}\mathrm{Re}\int_{\boldsymbol{x}\in\mathcal{S}_0}\left(\hat{\boldsymbol{E}}^{\mathrm{R}}\times\hat{\boldsymbol{H}}^{\mathrm{R}\circledast}\right)\cdot\boldsymbol{\nu}\,\mathrm{d}A$$

$$\tfrac{1}{2}\hat{P}^{\mathrm{s}}(\mathrm{i}\omega) = \tfrac{1}{2}\mathrm{Re}\int_{\boldsymbol{x}\in\mathcal{S}_0}\left(\hat{\boldsymbol{E}}^{\mathrm{s}}\times\hat{\boldsymbol{H}}^{\mathrm{s}\circledast}\right)\cdot\boldsymbol{\nu}\,\mathrm{d}A$$

as the time-averaged power absorbed and scattered by the receiving antenna system. Recall that in the limit $s = \delta + \mathrm{i}\omega$, $\delta \downarrow 0$, superscript $\circledast$ has the meaning of complex conjugate. The resulting relation is then normalized with respect to the magnitude of the (time-averaged) power-flow density of the incident plane wave that reads

$$\hat{\boldsymbol{S}}^{\mathrm{i}}(\mathrm{i}\omega) = \tfrac{1}{4}\left(\hat{\boldsymbol{E}}^{\mathrm{i}}\times\hat{\boldsymbol{H}}^{\mathrm{i}\circledast} + \hat{\boldsymbol{E}}^{\mathrm{i}\circledast}\times\hat{\boldsymbol{H}}^{\mathrm{i}}\right) = \tfrac{1}{2}\boldsymbol{\beta}\,\eta_0|\hat{e}^{\mathrm{i}}(\mathrm{i}\omega)|^2$$

and write

$$\frac{\hat{P}^{\mathrm{a}}(\mathrm{i}\omega)+\hat{P}^{\mathrm{s}}(\mathrm{i}\omega)}{2|\hat{\boldsymbol{S}}^{\mathrm{i}}(\mathrm{i}\omega)|} = -\frac{c_0}{\omega}\frac{\mathrm{Im}[e^{\mathrm{i}\circledast}(\mathrm{i}\omega)\boldsymbol{\alpha}\cdot\hat{\boldsymbol{E}}^{\mathrm{s};\infty}(\boldsymbol{\beta},\mathrm{i}\omega)]}{|\hat{e}^{\mathrm{i}}(\mathrm{i}\omega)|^2} \tag{3.19}$$

Note that the latter can be directly related to the extinction cross section

$$\hat{\sigma}^{\mathrm{c}}(\mathrm{i}\omega) = \hat{\sigma}^{\mathrm{a}}(\mathrm{i}\omega) + \hat{\sigma}^{\mathrm{s}}(\mathrm{i}\omega) \tag{3.20}$$

that is proportional to the total power extracted from the incident plane wave (see Ref. [46], Sec. 11.2).

- Express the antenna forward-scattering theorem in the time domain.

Hint: Using Eqs. (1.19) and (1.20) together with (1.14) evaluated at $t = 0$ (note the property (1.16)), we may express the total EM energy absorbed and scattered by the receiving antenna system:

$$W^{\mathrm{a}} = -\int_{\tau\in\mathbb{R}} \mathrm{d}\tau \int_{\boldsymbol{x}\in\mathcal{S}_0} \left[\boldsymbol{E}^{\mathrm{R}}(\boldsymbol{x},\tau)\times \boldsymbol{H}^{\mathrm{R}}(\boldsymbol{x},\tau)\right]\cdot \boldsymbol{\nu}\,\mathrm{d}A$$

$$W^{\mathrm{s}} = \int_{\tau\in\mathbb{R}} \mathrm{d}\tau \int_{\boldsymbol{x}\in\mathcal{S}_0} \left[\boldsymbol{E}^{\mathrm{s}}(\boldsymbol{x},\tau)\times \boldsymbol{H}^{\mathrm{s}}(\boldsymbol{x},\tau)\right]\cdot \boldsymbol{\nu}\,\mathrm{d}A \tag{3.21}$$

which allows rewriting Eq. (3.12) as

$$W^{\mathrm{a}} + W^{\mathrm{s}} = -\mu_0^{-1}\int_{\tau\in\mathbb{R}} \alpha e^{\mathrm{i}}(\tau)\cdot \partial_t^{-1}\boldsymbol{E}^{\mathrm{s};\infty}(\boldsymbol{\beta},\tau)\mathrm{d}\tau \tag{3.22}$$

where ∂_t^{-1} denotes the time integration operator. In the absorbed energy, we can further distinguish between the total EM energy dissipated in the antenna body, W^{h}, and in the antenna load. Hence, we write

$$W^{\mathrm{h}} + \sum_{n=1}^{N}\int_{\tau\in\mathbb{R}} V_n^{\mathrm{R}}(\tau)I_n^{\mathrm{R}}(\tau)\mathrm{d}\tau + W^{\mathrm{s}}$$

$$= -\mu_0^{-1}\int_{\tau\in\mathbb{R}} \alpha e^{\mathrm{i}}(\tau)\cdot \partial_t^{-1}\boldsymbol{E}^{\mathrm{s};\infty}(\boldsymbol{\beta},\tau)\mathrm{d}\tau \tag{3.23}$$

For the derivation based solely on space–time arguments that do not require the antenna system to be linear in its EM behavior, we refer the reader to Refs [13,48].

4

Antenna Matching Theorems

The maximum power transfer theorem in antenna theory is a corollary of the property of antenna self-reciprocity that makes possible to represent a receiving antenna system as a Kirchhoff-type equivalent network. This will be rigorously justified in Chapter 5 for an N-port antenna system with the aid of the reciprocity theorem of the time-convolution type applied to one and the same antenna system operating either as a transmitter or as a receiver (see Refs [12,19]). A classic result in this respect, known as the matching condition, states that for a given radiation $N \times N$ matrix impedance $\underline{\hat{Z}}^{\mathrm{T}}(s)$, the (time-averaged) power dissipated in the antenna load is maximized whenever the impedance $N \times N$ matrix of the load $\underline{\hat{Z}}^{\mathrm{L}}(s)$ is "matched" according to

$$\underline{\hat{Z}}^{\mathrm{L}}(\mathrm{i}\omega) = \underline{\hat{Z}}^{\mathrm{T}\circledast}(\mathrm{i}\omega) \tag{4.1}$$

for all $\omega \in \mathbb{R}$ of interest. Hence, the maximum power transfer is achieved if the impedance matrix of the load is complex conjugate to the corresponding radiation matrix impedance. Now, realizing the fact that the reciprocity theorem makes possible to couple the (local) Kirchhoff quantities of the antenna equivalent circuit to (global) EM wave field quantities in the antenna embedding, an interesting question arises what is the global-type counterpart of the matching condition (4.1). Introducing such a global-type matching condition for a class of reciprocal antennas is exactly the main purpose of this chapter.

4.1 Reciprocity Analysis of the Time-Correlation Type

The transmitting and receiving situations of an N-port antenna system (see Figure 4.1) are mutually interrelated using the reciprocity theorem of the time-correlation type.

4.1.1 Transmitting State

The N-port antenna system is in its transmitting (T) state excited by either electric feeding currents $\hat{I}_n^{\mathrm{T}}(s)$ or voltages $\hat{V}_n^{\mathrm{T}}(s)$ applied to its ports. These

Electromagnetic Reciprocity in Antenna Theory, First Edition. Martin Stumpf.

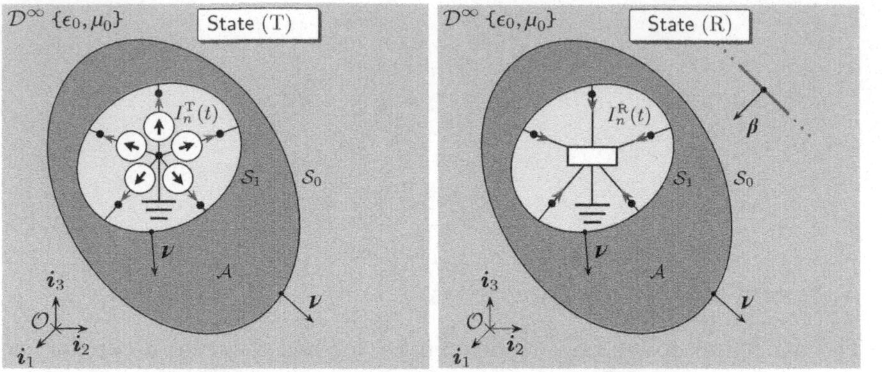

Figure 4.1 Transmitting (T) and receiving (R) states of the antenna system.

quantities are mutually interrelated via

$$\hat{V}^{\mathrm{T}}_m(s) = \sum_{n=1}^{N} \hat{Z}^{\mathrm{T}}_{m,n}(s)\hat{I}^{\mathrm{T}}_n(s) \tag{4.2}$$

for $m = \{1, \dots, N\}$, where $\hat{Z}^{\mathrm{T}}_{m,n}(s)$ is the matrix of the antenna radiation impedance, or, equivalently, via

$$\hat{I}^{\mathrm{T}}_n(s) = \sum_{m=1}^{N} \hat{Y}^{\mathrm{T}}_{n,m}(s)\hat{V}^{\mathrm{T}}_m(s) \tag{4.3}$$

for $n = \{1, \dots, N\}$, using the matrix of the antenna radiation admittance $\hat{Y}^{\mathrm{T}}_{n,m}(s)$. Combination of Eqs. (4.2) and (4.3) implies that the radiation impedance and admittance matrices are each other's inverse, that is,

$$\sum_{n=1}^{N} \hat{Z}^{\mathrm{T}}_{m,n}(s)\hat{Y}^{\mathrm{T}}_{n,p}(s) = \delta_{m,p} \tag{4.4}$$

for all $m = \{1, \dots, N\}$, $p = \{1, \dots, N\}$ and

$$\sum_{m=1}^{N} \hat{Y}^{\mathrm{T}}_{n,m}(s)\hat{Z}^{\mathrm{T}}_{m,q}(s) = \delta_{n,q} \tag{4.5}$$

for all $n = \{1, \dots, N\}$ and $q = \{1, \dots, N\}$. An important property of the impedance and admittance radiation matrices arises when the medium in $\mathcal{A}$ is self-adjoint in its EM properties. To demonstrate this property, the reciprocity theorem of the time-convolution type is applied to domain $\mathcal{A}$ and to two transmitting field states, say (T) and ($\tilde{\mathrm{T}}$), whose material states are each

Table 4.1 Application of the reciprocity theorem.

Domain $\mathcal{A}$		
Time-convolution	State (T)	State ($\tilde{\mathrm{T}}$)
Source	0	0
Field	$\{\hat{\boldsymbol{E}}^{\mathrm{T}}, \hat{\boldsymbol{H}}^{\mathrm{T}}\}$	$\{\hat{\boldsymbol{E}}^{\tilde{\mathrm{T}}}, \hat{\boldsymbol{H}}^{\tilde{\mathrm{T}}}\}$
Material	$\{\underline{\hat{\eta}}, \underline{\hat{\zeta}}\}$	$\{\underline{\hat{\eta}}^{T}, \underline{\hat{\zeta}}^{T}\}$

other's adjoint (see Table 4.1). In this way, making use of Eq. (1.30), we get

$$\int_{\boldsymbol{x}\in\mathcal{S}_0} \left(\hat{\boldsymbol{E}}^{\mathrm{T}} \times \hat{\boldsymbol{H}}^{\tilde{\mathrm{T}}} - \hat{\boldsymbol{E}}^{\tilde{\mathrm{T}}} \times \hat{\boldsymbol{H}}^{\mathrm{T}}\right) \cdot \boldsymbol{\nu}\,\mathrm{d}A = \int_{\boldsymbol{x}\in\mathcal{S}_1} \left(\hat{\boldsymbol{E}}^{\mathrm{T}} \times \hat{\boldsymbol{H}}^{\tilde{\mathrm{T}}} - \hat{\boldsymbol{E}}^{\tilde{\mathrm{T}}} \times \hat{\boldsymbol{H}}^{\mathrm{T}}\right) \cdot \boldsymbol{\nu}\,\mathrm{d}A \tag{4.6}$$

Since both field states are causal and the embedding is self-adjoint in its EM constitutive properties, we can make use of Eq. (1.44) and write

$$\int_{\boldsymbol{x}\in\mathcal{S}_0} \left(\hat{\boldsymbol{E}}^{\mathrm{T}} \times \hat{\boldsymbol{H}}^{\tilde{\mathrm{T}}} - \hat{\boldsymbol{E}}^{\tilde{\mathrm{T}}} \times \hat{\boldsymbol{H}}^{\mathrm{T}}\right) \cdot \boldsymbol{\nu}\,\mathrm{d}A = 0 \tag{4.7}$$

while the integral over the terminal surface reads

$$\int_{\boldsymbol{x}\in\mathcal{S}_1} \left(\hat{\boldsymbol{E}}^{\mathrm{T}} \times \hat{\boldsymbol{H}}^{\tilde{\mathrm{T}}} - \hat{\boldsymbol{E}}^{\tilde{\mathrm{T}}} \times \hat{\boldsymbol{H}}^{\mathrm{T}}\right) \cdot \boldsymbol{\nu}\,\mathrm{d}A \simeq \sum_{n=1}^{N} [\hat{V}_n^{\mathrm{T}}(s)\hat{I}_n^{\tilde{\mathrm{T}}}(s) - \hat{V}_n^{\tilde{\mathrm{T}}}(s)\hat{I}_n^{\mathrm{T}}(s)] \tag{4.8}$$

where we have taken into account the orientation of the electric currents with respect to the outer unit vector $\boldsymbol{\nu}$ along $\mathcal{S}_1$. Finally, combination of Eqs. (4.6)–(4.8) with Eqs. (4.2) and (4.3) yields

$$\{\hat{Z}^{\mathrm{T}}_{n,m}, \hat{Y}^{\mathrm{T}}_{m,n}\}(s) = \{\hat{Z}^{\tilde{\mathrm{T}}}_{m,n}, \hat{Y}^{\tilde{\mathrm{T}}}_{n,m}\}(s) \tag{4.9}$$

for all $m = \{1, \ldots, N\}, n = \{1, \ldots, N\}$. The total radiated EM wave field is causal and hence admits the far-field expansion ([16], Sec. 26.11):

$$\{\hat{\boldsymbol{E}}^{\mathrm{T}}, \hat{\boldsymbol{H}}^{\mathrm{T}}\}(\boldsymbol{x}, s) = \{\hat{\boldsymbol{E}}^{\mathrm{T};\infty}, \hat{\boldsymbol{H}}^{\mathrm{T};\infty}\}(\boldsymbol{\xi}, s) \exp(-s|\boldsymbol{x}|/c_0)(4\pi|\boldsymbol{x}|)^{-1}\left[1 + O(|\boldsymbol{x}|^{-1})\right] \tag{4.10}$$

as $|\boldsymbol{x}| \to \infty$, with $\hat{\boldsymbol{H}}^{\mathrm{T};\infty} = \eta_0\boldsymbol{\xi} \times \hat{\boldsymbol{E}}^{\mathrm{T};\infty}$ and $\boldsymbol{\xi} \cdot \hat{\boldsymbol{E}}^{\mathrm{T};\infty} = 0$. This allows, using Huygens' surface-integral representation of the radiated field, expressing the far-field amplitude in terms of the equivalent electric and magnetic current surface densities on the bounding surface of the antenna system (cf. Ref. [11], Eq. (3.8)):

$$\hat{\boldsymbol{E}}^{\mathrm{T};\infty}(\boldsymbol{\xi}, s) = s\mu_0(\boldsymbol{\xi}\boldsymbol{\xi} - \underline{\boldsymbol{I}}) \int_{\boldsymbol{x}\in\mathcal{S}_0} \exp(s\boldsymbol{\xi} \cdot \boldsymbol{x}/c_0)\left[\boldsymbol{\nu}(\boldsymbol{x}) \times \hat{\boldsymbol{H}}^{\mathrm{T}}(\boldsymbol{x}, s)\right]\mathrm{d}A$$

$$+ sc_0^{-1}\boldsymbol{\xi} \times \int_{\boldsymbol{x}\in\mathcal{S}_0} \exp(s\boldsymbol{\xi} \cdot \boldsymbol{x}/c_0)\left[\hat{\boldsymbol{E}}^{\mathrm{T}}(\boldsymbol{x}, s) \times \boldsymbol{\nu}(\boldsymbol{x})\right]\mathrm{d}A \tag{4.11}$$

for all $\boldsymbol{\xi} \in \Omega = \{\boldsymbol{\xi} \cdot \boldsymbol{\xi} = 1\}$ on the unit sphere.

4.1.2 Receiving State

The N-port antenna system is in its receiving state irradiated by the uniform EM plane wave:

$$\hat{\boldsymbol{E}}^{\mathrm{i}}(\boldsymbol{x}, s) = \boldsymbol{\alpha}\,\hat{e}^{\mathrm{i}}(s)\exp(-s\boldsymbol{\beta} \cdot \boldsymbol{x}/c_0) \qquad \text{(3.1 revisited)}$$

$$\hat{\boldsymbol{H}}^{\mathrm{i}}(\boldsymbol{x}, s) = (\boldsymbol{\beta} \times \boldsymbol{\alpha})\,\eta_0\hat{e}^{\mathrm{i}}(s)\exp(-s\boldsymbol{\beta} \cdot \boldsymbol{x}/c_0) \qquad \text{(3.2 revisited)}$$

In a standard way, we next introduce the scattered EM wave field that accounts for the presence of the receiving antenna system as the difference between the total EM field in the configuration, $\{\hat{\boldsymbol{E}}^{\mathrm{R}}, \hat{\boldsymbol{H}}^{\mathrm{R}}\}$, and the incident wave field, namely

$$\{\hat{\boldsymbol{E}}^{\mathrm{s}}, \hat{\boldsymbol{H}}^{\mathrm{s}}\}(\boldsymbol{x}, s) \triangleq \{\hat{\boldsymbol{E}}^{\mathrm{R}}, \hat{\boldsymbol{H}}^{\mathrm{R}}\}(\boldsymbol{x}, s) - \{\hat{\boldsymbol{E}}^{\mathrm{i}}, \hat{\boldsymbol{H}}^{\mathrm{i}}\}(\boldsymbol{x}, s) \qquad \text{(3.3 revisited)}$$

The scattered field is causal and hence has, far enough from the antenna system, the form of a spherical wave expanding away from the origin, with $\{\hat{\boldsymbol{E}}^{\mathrm{s};\infty}, \hat{\boldsymbol{H}}^{\mathrm{s};\infty}\}$ being its far-field amplitudes (see Eq. (3.4)). Consequently, the induced voltage across the port of the receiving antenna system can be expressed as

$$\hat{V}_m^{\mathrm{R}}(s) = \sum_{n=1}^{N} \hat{Z}_{m,n}^{\mathrm{L}}(s)\hat{I}_n^{\mathrm{R}}(s) \tag{4.12}$$

for $m = \{1, \ldots, N\}$, where $\hat{Z}_{m,n}^{\mathrm{L}}(s)$ is the impedance matrix of the antenna load. Equivalently, we may write

$$\hat{I}_n^{\mathrm{R}}(s) = \sum_{m=1}^{N} \hat{Y}_{n,m}^{\mathrm{L}}(s)\hat{V}_m^{\mathrm{R}}(s) \tag{4.13}$$

for $n = \{1, \ldots, N\}$, using the admittance matrix of the antenna load $\hat{Y}_{n,m}^{\mathrm{L}}(s)$. Combination of Eqs. (4.12) and (4.13) implies that the impedance and admittance matrices of the antenna load are each other's inverse. Accordingly, these

Table 4.2 Application of the reciprocity theorem.

Domain $\mathcal{D}^{\infty}$		
Time-correlation	State (s)	State (T)
Source	0	0
Field	$\{\hat{\boldsymbol{E}}^{\mathrm{s}}, \hat{\boldsymbol{H}}^{\mathrm{s}}\}$	$\{\hat{\boldsymbol{E}}^{\mathrm{T}}, \hat{\boldsymbol{H}}^{\mathrm{T}}\}$
Material	$\{\epsilon_0, \mu_0\}$	$\{\epsilon_0, \mu_0\}$

load matrices are interrelated in the same way as the radiation impedance and admittance matrices (see Eqs. (4.4) and (4.5)).

4.1.3 Equivalent Matching Condition

In our efforts to find the global-type counterpart of the classic (conjugate) matching condition (4.1), we start with the combination of Eqs. (3.1) and (3.2) and (4.11) that yields

$$\int_{\boldsymbol{x}\in\mathcal{S}_0} \left(\hat{\boldsymbol{E}}^{\mathrm{i}} \times \hat{\boldsymbol{H}}^{\mathrm{T}\circledast} + \hat{\boldsymbol{E}}^{\mathrm{T}\circledast} \times \hat{\boldsymbol{H}}^{\mathrm{i}}\right) \cdot \boldsymbol{\nu}\,\mathrm{d}A$$
$$= -\hat{e}^{\mathrm{i}}(s)\,\boldsymbol{\alpha} \cdot \hat{\boldsymbol{E}}^{\mathrm{T};\infty}(\boldsymbol{\beta}, -s)/s\mu_0 \tag{4.14}$$

In the next step, we will apply the reciprocity theorem of the time-correlation type to the domain exterior to the antenna system and to the scattered and transmitted EM wave fields (see Table 4.2). Owing to the fact that both of these wave fields are causal, we can directly make use of (1.45) and write

$$\int_{\boldsymbol{x}\in\mathcal{S}_0} \left(\hat{\boldsymbol{E}}^{\mathrm{s}} \times \hat{\boldsymbol{H}}^{\mathrm{T}\circledast} + \hat{\boldsymbol{E}}^{\mathrm{T}\circledast} \times \hat{\boldsymbol{H}}^{\mathrm{s}}\right) \cdot \boldsymbol{\nu}\,\mathrm{d}A$$
$$= \left(\eta_0/8\pi^2\right) \int_{\boldsymbol{\xi}\in\Omega} \hat{\boldsymbol{E}}^{\mathrm{s};\infty}(\boldsymbol{\xi}, s) \cdot \hat{\boldsymbol{E}}^{\mathrm{T};\infty}(\boldsymbol{\xi}, -s)\mathrm{d}\Omega \tag{4.15}$$

The sum of Eqs. (4.14) and (4.15) along with Eq. (3.3) results in the following reciprocity relation:

$$\int_{\boldsymbol{x}\in\mathcal{S}_0} \left(\hat{\boldsymbol{E}}^{\mathrm{R}} \times \hat{\boldsymbol{H}}^{\mathrm{T}\circledast} + \hat{\boldsymbol{E}}^{\mathrm{T}\circledast} \times \hat{\boldsymbol{H}}^{\mathrm{R}}\right) \cdot \boldsymbol{\nu}\,\mathrm{d}A$$
$$= \left(\eta_0/8\pi^2\right) \int_{\boldsymbol{\xi}\in\Omega} \hat{\boldsymbol{E}}^{\mathrm{s};\infty}(\boldsymbol{\xi}, s) \cdot \hat{\boldsymbol{E}}^{\mathrm{T};\infty}(\boldsymbol{\xi}, -s)\mathrm{d}\Omega$$
$$- \hat{e}^{\mathrm{i}}(s)\,\boldsymbol{\alpha} \cdot \hat{\boldsymbol{E}}^{\mathrm{T};\infty}(\boldsymbol{\beta}, -s)/s\mu_0 \tag{4.16}$$

Now, applying the reciprocity theorem of the time-correlation type to the domain occupied by the antenna and to the total EM wave fields in the transmitting and receiving situations (see Table 4.3) yields

$$\begin{aligned}
&\int_{\boldsymbol{x}\in\mathcal{S}_0} \left(\hat{\boldsymbol{E}}^{\mathrm{R}} \times \hat{\boldsymbol{H}}^{\mathrm{T}\circledast} + \hat{\boldsymbol{E}}^{\mathrm{T}\circledast} \times \hat{\boldsymbol{H}}^{\mathrm{R}}\right) \cdot \boldsymbol{\nu}\,\mathrm{d}A \\
&= \int_{\boldsymbol{x}\in\mathcal{S}_1} \left(\hat{\boldsymbol{E}}^{\mathrm{R}} \times \hat{\boldsymbol{H}}^{\mathrm{T}\circledast} + \hat{\boldsymbol{E}}^{\mathrm{T}\circledast} \times \hat{\boldsymbol{H}}^{\mathrm{R}}\right) \cdot \boldsymbol{\nu}\,\mathrm{d}A \\
&\qquad - \int_{\boldsymbol{x}\in\mathcal{A}} \left\{ \hat{\boldsymbol{H}}^{\mathrm{R}} \cdot \left[\underline{\hat{\zeta}}^{\circledast} + \underline{\hat{\zeta}}^{T}\right] \cdot \hat{\boldsymbol{H}}^{\mathrm{T}\circledast} \right. \\
&\qquad\qquad \left. + \hat{\boldsymbol{E}}^{\mathrm{R}} \cdot \left[\underline{\hat{\eta}}^{\circledast} + \underline{\hat{\eta}}^{T}\right] \cdot \hat{\boldsymbol{E}}^{\mathrm{T}\circledast} \right\} \mathrm{d}V
\end{aligned} \tag{4.17}$$

where the volume integral vanishes for the medium that is time-reverse self-adjoint in its EM properties (see Eqs. (1.37) and (1.38)). This includes loss-free antenna systems such as instantaneously reacting antennas described with symmetric tensors of electric permittivity and magnetic permeability (see Eqs. (1.48) and (1.49)), antenna systems made from PEC components (e.g., Figure 1.7), or, eventually, the combination of these cases. Since the surface integral over the terminal surface of the antenna system can be expressed in terms of Kirchhoff quantities, that is,

$$\begin{aligned}
&\int_{\boldsymbol{x}\in\mathcal{S}_1} \left(\hat{\boldsymbol{E}}^{\mathrm{R}} \times \hat{\boldsymbol{H}}^{\mathrm{T}\circledast} + \hat{\boldsymbol{E}}^{\mathrm{T}\circledast} \times \hat{\boldsymbol{H}}^{\mathrm{R}}\right) \cdot \boldsymbol{\nu}\,\mathrm{d}A \\
&\qquad \simeq \sum_{m=1}^{N} \left[\hat{V}_m^{\mathrm{R}}(s)\hat{I}_m^{\mathrm{T}}(-s) - \hat{V}_m^{\mathrm{T}}(-s)\hat{I}_m^{\mathrm{R}}(s)\right]
\end{aligned} \tag{4.18}$$

we may, upon combining Eqs. (4.16)–(4.18), write

$$\begin{aligned}
&\frac{\eta_0}{8\pi^2} \int_{\boldsymbol{\xi}\in\Omega} \hat{\boldsymbol{E}}^{\mathrm{s};\infty}(\boldsymbol{\xi}, s) \cdot \hat{\boldsymbol{E}}^{\mathrm{T};\infty}(\boldsymbol{\xi}, -s)\mathrm{d}\Omega - \frac{1}{s\mu_0}\hat{e}^{\mathrm{i}}(s)\,\boldsymbol{\alpha} \cdot \hat{\boldsymbol{E}}^{\mathrm{T};\infty}(\boldsymbol{\beta}, -s) \\
&= \sum_{m=1}^{N} \left[\hat{V}_m^{\mathrm{R}}(s)\hat{I}_m^{\mathrm{T}}(-s) - \hat{V}_m^{\mathrm{T}}(-s)\hat{I}_m^{\mathrm{R}}(s)\right]
\end{aligned} \tag{4.19}$$

for loss-free antennas. Making use of Eqs. (4.2) and (4.12) in the right-hand side of Eq. (4.19), it is noted that

$$\begin{aligned}
&\sum_{m=1}^{N} \left[\hat{V}_m^{\mathrm{R}}(s)\hat{I}_m^{\mathrm{T}}(-s) - \hat{V}_m^{\mathrm{T}}(-s)\hat{I}_m^{\mathrm{R}}(s)\right] \\
&= \sum_{m=1}^{N}\sum_{n=1}^{N} \left[\hat{Z}_{m,n}^{\mathrm{L}}(s) - \hat{Z}_{n,m}^{\mathrm{T}}(-s)\right] \hat{I}_n^{\mathrm{R}}(s)\hat{I}_m^{\mathrm{T}}(-s)
\end{aligned} \tag{4.20}$$

Table 4.3 Application of the reciprocity theorem.

Domain $\mathcal{A}$		
Time-correlation	State (R)	State (T)
Source	0	0
Field	$\{\hat{\boldsymbol{E}}^{\mathrm{R}}, \hat{\boldsymbol{H}}^{\mathrm{R}}\}$	$\{\hat{\boldsymbol{E}}^{\mathrm{T}}, \hat{\boldsymbol{H}}^{\mathrm{T}}\}$
Material	$\{\hat{\underline{\eta}}, \hat{\underline{\zeta}}\}$	$\{\hat{\underline{\eta}}, \hat{\underline{\zeta}}\}$

is zero provided that the local-type matching condition (4.1) applies and that the antenna system is reciprocal in its EM behavior (see Eq. (4.9)). This, with regard to Eq. (4.19), implies that the matching condition (4.1) is for *reciprocal* antennas equivalent to stating that

$$(\eta_0/8\pi^2) \int_{\xi\in\Omega} \hat{\boldsymbol{E}}^{\mathrm{S};\infty}(\boldsymbol{\xi}, s) \cdot \hat{\boldsymbol{E}}^{\mathrm{T};\infty}(\boldsymbol{\xi}, -s)\mathrm{d}\Omega = \hat{e}^{\mathrm{i}}(s)\,\boldsymbol{\alpha} \cdot \hat{\boldsymbol{E}}^{\mathrm{T};\infty}(\boldsymbol{\beta}, -s)/s\mu_0 \tag{4.21}$$

for all $\boldsymbol{\beta} \in \Omega$. The latter condition can be further rewritten with the help of the far-field scattering tensor of the electric type, $\hat{\underline{\boldsymbol{S}}}(\boldsymbol{\xi}|\boldsymbol{\beta}, s)$, and the effective length of the current-excited transmitting antenna system, $\hat{\boldsymbol{\ell}}_m^{\mathrm{T;I}}$, namely,

$$\hat{\boldsymbol{E}}^{\mathrm{S};\infty}(\boldsymbol{\xi}, s) = \hat{\underline{\boldsymbol{S}}}(\boldsymbol{\xi}|\boldsymbol{\beta}, s) \cdot \boldsymbol{\alpha}\hat{e}^{\mathrm{i}}(s) \tag{4.22}$$

$$\hat{\boldsymbol{E}}^{\mathrm{T};\infty}(\boldsymbol{\xi}, s) = s\mu_0 \sum_{m=1}^{N} \hat{I}_m^{\mathrm{T}}(s)\hat{\boldsymbol{\ell}}_m^{\mathrm{T;I}}(\boldsymbol{\xi}, s) \tag{4.23}$$

Substitution of Eqs. (4.22) and (4.23) into Eq. (4.21) gives

$$\int_{\xi\in\Omega} \hat{\boldsymbol{\ell}}_m^{\mathrm{T;I}}(\boldsymbol{\xi}, -s) \cdot \hat{\underline{\boldsymbol{S}}}(\boldsymbol{\xi}|\boldsymbol{\beta}, s) \cdot \boldsymbol{\alpha}\,\mathrm{d}\Omega = \frac{8\pi^2 c_0}{s}\boldsymbol{\alpha} \cdot \hat{\boldsymbol{\ell}}_m^{\mathrm{T;I}}(\boldsymbol{\beta}, -s) \tag{4.24}$$

for all $m = \{1, \ldots, N\}$. A special case of the global-type matching condition concerning a 1-port antenna system can be found in Ref. [43]

Exercises

- Verify the validity of the global-type matching condition for a short wire carrying a uniform electric current.

Hint: Start from the surface integral representation for the scattered field (3.5), apply the explicit-type boundary condition on the PEC surface of the wire dipole, and take the limit $s \downarrow 0$. In this way, the electric dipole contribution to the scattering tensor $\underline{\hat{\boldsymbol{S}}}$ can be expressed as

$$\underline{\hat{\boldsymbol{S}}}^{\mathrm{E}}(\boldsymbol{\xi}|\boldsymbol{\beta}, s) = -(s^2/c_0^2)(\underline{\boldsymbol{I}} - \boldsymbol{\xi}\boldsymbol{\xi}) \cdot \underline{\hat{\boldsymbol{\alpha}}}^{\mathrm{E}} \cdot (\underline{\boldsymbol{I}} - \boldsymbol{\beta}\boldsymbol{\beta})$$

in which $\underline{\hat{\boldsymbol{\alpha}}}^{\mathrm{E}}$ is the electric polarizability tensor that is related to the electric dipole moment via

$$\hat{\boldsymbol{p}}(s) = \epsilon_0 \hat{e}^{\mathrm{i}}(s)\, \underline{\hat{\boldsymbol{\alpha}}}^{\mathrm{E}} \cdot \boldsymbol{\alpha}$$

Subsequently, express the electric dipole moment in terms of the induced electric current flowing across the antenna load, that is,

$$\hat{\boldsymbol{p}}(s) = -s^{-1} \hat{I}^{\mathrm{R}}(s)\, \boldsymbol{\ell}$$

where $\boldsymbol{\ell}$ is the vectorial length of the wire. The electric current $\hat{I}^{\mathrm{R}}$ can be found with the aid of Thévenin's equivalent representation of a 1-port antenna system. The open-circuit voltage generator for the incident plane wave can be expressed in closed form ([18], Eq. (56))

$$\hat{V}^{\mathrm{G}}(s) = \hat{e}^{\mathrm{i}}(s)\boldsymbol{\alpha} \cdot \hat{\boldsymbol{E}}^{\mathrm{T};\infty}(-\boldsymbol{\beta}, s)/s\mu_0 \hat{I}^{\mathrm{T}}(s)$$

where the radiation amplitude of a conducting wire carrying a uniform electric current distribution is given as

$$\hat{\boldsymbol{E}}^{\mathrm{T};\infty}(\boldsymbol{\xi}, s) = s\mu_0 \hat{I}^{\mathrm{T}}(s)(\boldsymbol{\xi}\boldsymbol{\xi} - \underline{\boldsymbol{I}}) \cdot \boldsymbol{\ell}$$

and the corresponding effective length follows (see Eq. (4.23)):

$$\hat{\boldsymbol{\ell}}^{\mathrm{T;I}}(\boldsymbol{\xi}, s) = (\boldsymbol{\xi}\boldsymbol{\xi} - \underline{\boldsymbol{I}}) \cdot \boldsymbol{\ell}$$

Hence, upon collecting the results, the polarizability tensor of the electric dipole can be expressed as

$$\underline{\hat{\boldsymbol{\alpha}}}^{\mathrm{E}} = (s\epsilon_0)^{-1}(\hat{Z}^{\mathrm{T}} + \hat{Z}^{\mathrm{L}})^{-1}\, \boldsymbol{\ell}\, \boldsymbol{\ell}$$

where we have used $\hat{I}^{\mathrm{R}} = \hat{V}^{\mathrm{G}}/(\hat{Z}^{\mathrm{T}} + \hat{Z}^{\mathrm{L}})$. Knowing the polarizability tensor, the corresponding scattering tensor follows:

$$\underline{\hat{\boldsymbol{S}}}^{\mathrm{E}}(\boldsymbol{\xi}|\boldsymbol{\beta}, s) = -s\mu_0(\hat{Z}^{\mathrm{T}} + \hat{Z}^{\mathrm{L}})^{-1}[(\underline{\boldsymbol{I}} - \boldsymbol{\xi}\boldsymbol{\xi}) \cdot \boldsymbol{\ell}][\boldsymbol{\ell} \cdot (\underline{\boldsymbol{I}} - \boldsymbol{\beta}\boldsymbol{\beta})]$$

With the expressions for the scattering tensor and the effective length at our disposal, we may evaluate the integral in Eq. (4.24) and get

$$\int_{\boldsymbol{\xi}\in\Omega} \hat{\boldsymbol{\ell}}^{\mathrm{T;I}}(\boldsymbol{\xi}, -s) \cdot \underline{\hat{\boldsymbol{S}}}^{\mathrm{E}}(\boldsymbol{\xi}|\boldsymbol{\beta}, s) \cdot \boldsymbol{\alpha}\, \mathrm{d}\Omega = \frac{8\pi}{3} \frac{s\mu_0\ell^2}{\hat{Z}^{\mathrm{T}} + \hat{Z}^{\mathrm{L}}} \boldsymbol{\alpha} \cdot \boldsymbol{\ell} \tag{4.25}$$

where $\ell = |\boldsymbol{\ell}|$ is the length of the wire. Substitution of the results in the global-type matching condition (4.24) results in

$$\hat{Z}^{\mathrm{T}}(s) + \hat{Z}^{\mathrm{L}}(s) = -s^2\mu_0\ell^2/3\pi c_0 \tag{4.26}$$

where we have used $\boldsymbol{\alpha} \cdot \hat{\boldsymbol{\ell}}^{\mathrm{T;I}}(\boldsymbol{\beta}, -s) = -\boldsymbol{\alpha} \cdot \boldsymbol{\ell}$ (see also Exercises in Chapter 9). The last equality becomes more transparent in the real-FD under the limit $\{s = \delta + \mathrm{i}\omega, \delta \downarrow 0, \omega \in \mathbb{R}\}$, that is,

$$\hat{Z}^{\mathrm{T}}(\mathrm{i}\omega) + \hat{Z}^{\mathrm{L}}(\mathrm{i}\omega) = (\mu_0/\epsilon_0)^{1/2}(4\pi/3)(\ell/\lambda)^2 \tag{4.27}$$

where $\lambda = 2\pi c_0/\omega \in \mathbb{R}$ with $c_0 = (\epsilon_0\mu_0)^{-1/2}$. The right-hand side of the last equation can be identified with the radiation resistance ([30], Sec. 4.4a, Eq. (10)) of a short-wire antenna and we get

$$\hat{Z}^{\mathrm{T}}(\mathrm{i}\omega) + \hat{Z}^{\mathrm{L}}(\mathrm{i}\omega) = 2\mathrm{Re}[\hat{Z}^{\mathrm{T}}(\mathrm{i}\omega)] \tag{4.28}$$

which is obviously satisfied when the classic (conjugate) matching condition (4.1) applies.

- Verify the validity of the global-type matching condition for a small loop carrying a uniform electric current.

Hint: The solution follows the lines of reasoning from the previous example. At first, express the magnetic dipole contribution to the scattering tensor $\underline{\hat{\boldsymbol{S}}}$ as

$$\underline{\hat{\boldsymbol{S}}}^{\mathrm{H}}(\boldsymbol{\xi}|\boldsymbol{\beta}, s) = (s^2/c_0^2)\, \boldsymbol{\xi} \times \underline{\hat{\boldsymbol{\alpha}}}^{\mathrm{H}} \times \boldsymbol{\beta}$$

in which $\underline{\hat{\boldsymbol{\alpha}}}^{\mathrm{H}}$ is the magnetic polarizability tensor that is related to the magnetic dipole moment via

$$\hat{\boldsymbol{m}}(s) = (\epsilon_0/\mu_0)^{1/2}\hat{e}^{\mathrm{i}}(s)\, \underline{\hat{\boldsymbol{\alpha}}}^{\mathrm{H}} \cdot (\boldsymbol{\beta} \times \boldsymbol{\alpha})$$

Subsequently, express the magnetic dipole moment in terms of the induced electric current flowing across the antenna load, specifically

$$\hat{\boldsymbol{m}}(s) = -\hat{I}^{\mathrm{R}}(s)\,\boldsymbol{\mathcal{A}}$$

where $\boldsymbol{\mathcal{A}}$ is the vectorial area of the loop. As in the previous example, the electric current in the load is found using the reciprocity-based expression for the voltage generator in Thévenin's circuit representation ([18], Eq. (56)). In this expression, use the radiation characteristics of a conducting loop carrying a uniform electric current distribution, namely,

$$\hat{\boldsymbol{E}}^{\mathrm{T};\infty}(\boldsymbol{\xi}, s) = \mu_0 c_0^{-1} s^2 \hat{I}^{\mathrm{T}}(s)\,\boldsymbol{\xi} \times \boldsymbol{\mathcal{A}}$$

and the corresponding effective length follows (see Eq. (4.23)):

$$\hat{\boldsymbol{\ell}}^{\mathrm{T;I}}(\boldsymbol{\xi}, s) = s c_0^{-1}\,\boldsymbol{\xi} \times \boldsymbol{\mathcal{A}}$$

Hence, upon collecting the results, the polarizability tensor of the magnetic dipole can be expressed as

$$\underline{\hat{\boldsymbol{\alpha}}}^{\mathrm{H}} = -s\mu_0(\hat{Z}^{\mathrm{T}} + \hat{Z}^{\mathrm{L}})^{-1}\,\boldsymbol{\mathcal{A}}\,\boldsymbol{\mathcal{A}}$$

where we have used, again, $\hat{I}^{\mathrm{R}} = \hat{V}^{\mathrm{G}}/(\hat{Z}^{\mathrm{T}} + \hat{Z}^{\mathrm{L}})$. Knowing the polarizability tensor, the corresponding scattering tensor follows:

$$\underline{\hat{\boldsymbol{S}}}^{\mathrm{H}}(\boldsymbol{\xi}|\boldsymbol{\beta}, s) = (s^2/c_0^2)s\mu_0(\hat{Z}^{\mathrm{T}} + \hat{Z}^{\mathrm{L}})^{-1}(\boldsymbol{\xi} \times \boldsymbol{\mathcal{A}})(\boldsymbol{\beta} \times \boldsymbol{\mathcal{A}})$$

With the expressions for the scattering tensor and the effective length at our disposal, we may evaluate the integral in Eq. (4.24) and get

$$\int_{\boldsymbol{\xi}\in\Omega} \hat{\boldsymbol{\ell}}^{\mathrm{T;I}}(\boldsymbol{\xi}, -s) \cdot \underline{\hat{\boldsymbol{S}}}^{\mathrm{H}}(\boldsymbol{\xi}|\boldsymbol{\beta}, s) \cdot \boldsymbol{\alpha}\,\mathrm{d}\Omega = -\frac{8\pi}{3}\frac{s^3}{c_0^3}\frac{s\mu_0\mathcal{A}^2}{\hat{Z}^{\mathrm{T}} + \hat{Z}^{\mathrm{L}}}\,\boldsymbol{\alpha} \cdot (\boldsymbol{\beta} \times \boldsymbol{\mathcal{A}}) \tag{4.29}$$

where $\mathcal{A} = |\boldsymbol{\mathcal{A}}| = \pi r^2$ is the area of the loop. Substitution of the results in the global-type matching condition (4.24) results in

$$\hat{Z}^{\mathrm{T}}(s) + \hat{Z}^{\mathrm{L}}(s) = s^4\mu_0\mathcal{A}^2/3\pi c_0^3 \tag{4.30}$$

Again, the limit $\{s = \delta + \mathrm{i}\omega, \delta \downarrow 0, \omega \in \mathbb{R}\}$ reveals

$$\hat{Z}^{\mathrm{T}}(\mathrm{i}\omega) + \hat{Z}^{\mathrm{L}}(\mathrm{i}\omega) = (\mu_0/\epsilon_0)^{1/2}(16\pi^5/3)(r/\lambda)^2 \tag{4.31}$$

The right-hand side of the latter is twice the radiation resistance of a small current loop ([21], Sec. 2.6, Eq. (2.33)), which proves the validity of the global-type matching condition Eq. (4.24) for the small-loop antenna.

5

Equivalent Kirchhoff Network Representations of a Receiving Antenna System

The reciprocity relation between the transmit and receive antenna patterns is among the most famous properties in antenna theory. Another closely related consequence of EM reciprocity is the equivalent circuit representation of a receiving antenna system. In this chapter, therefore, the reciprocity theorem of the time-convolution type is systematically applied to construct Kirchhoff-type equivalent network representations of an N-port receiving antenna system. In accordance with Ref. [11], the Lorentz reciprocity theorem is applied to one and the same antenna system operating either in its transmitting or receiving state. The resulting (self-)reciprocity relation immediately offers the description for both the internal Thévenin-voltage and Norton-current generators. Closed-form expressions describing the internal generators are given for the antenna excitation via a plane wave and via a given volume current distribution. The description that follows is largely based on a seminal paper by de Hoop et al. [19], where the reciprocity analysis is carried out entirely in the time domain.

5.1 Reciprocity Analysis of the Time-Convolution Type

The transmitting and receiving situations of an N-port antenna system (see Figure 4.1) are mutually interrelated using the reciprocity theorem of the time-convolution type. The internal source strengths of the Kirchhoff equivalent networks are specified for both the EM plane-wave incidence and the known volume-source distribution.

5.1.1 Equivalent Circuits for Plane-Wave Incidence

To construct the Kirchhoff-type equivalent circuit for the antenna system that is irradiated by the uniform plane wave, namely,

Electromagnetic Reciprocity in Antenna Theory, First Edition. Martin Stumpf.

Table 5.1 Application of the reciprocity theorem.

Domain $\mathcal{D}^\infty$		
Time-convolution	State (s)	State (T)
Source	0	0
Field	$\{\hat{\boldsymbol{E}}^{\mathrm{s}}, \hat{\boldsymbol{H}}^{\mathrm{s}}\}$	$\{\hat{\boldsymbol{E}}^{\mathrm{T}}, \hat{\boldsymbol{H}}^{\mathrm{T}}\}$
Material	$\{\epsilon_0, \mu_0\}$	$\{\epsilon_0, \mu_0\}$

$$\hat{\boldsymbol{E}}^{\mathrm{i}}(\boldsymbol{x}, s) = \boldsymbol{\alpha}\, \hat{e}^{\mathrm{i}}(s) \exp(-s\boldsymbol{\beta} \cdot \boldsymbol{x}/c_0) \tag{3.1 revisited}$$

$$\hat{\boldsymbol{H}}^{\mathrm{i}}(\boldsymbol{x}, s) = (\boldsymbol{\beta} \times \boldsymbol{\alpha})\, \eta_0 \hat{e}^{\mathrm{i}}(s) \exp(-s\boldsymbol{\beta} \cdot \boldsymbol{x}/c_0) \tag{3.2 revisited}$$

we employ the surface integral representation of the radiated far-field amplitude,

$$\hat{\boldsymbol{E}}^{\mathrm{T};\infty}(\boldsymbol{\xi}, s) = s\mu_0(\boldsymbol{\xi}\boldsymbol{\xi} - \underline{\boldsymbol{I}}) \int_{\boldsymbol{x}\in\mathcal{S}_0} \exp(s\boldsymbol{\xi} \cdot \boldsymbol{x}/c_0)\left[\boldsymbol{\nu}(\boldsymbol{x}) \times \hat{\boldsymbol{H}}^{\mathrm{T}}(\boldsymbol{x}, s)\right] \mathrm{d}A$$
$$+ sc_0^{-1}\boldsymbol{\xi} \times \int_{\boldsymbol{x}\in\mathcal{S}_0} \exp(s\boldsymbol{\xi} \cdot \boldsymbol{x}/c_0)\left[\hat{\boldsymbol{E}}^{\mathrm{T}}(\boldsymbol{x}, s) \times \boldsymbol{\nu}(\boldsymbol{x})\right] \mathrm{d}A \tag{4.11 revisited}$$

and find that the following equality applies:

$$\int_{\boldsymbol{x}\in\mathcal{S}_0} \left(\hat{\boldsymbol{E}}^{\mathrm{i}} \times \hat{\boldsymbol{H}}^{\mathrm{T}} - \hat{\boldsymbol{E}}^{\mathrm{T}} \times \hat{\boldsymbol{H}}^{\mathrm{i}}\right) \cdot \boldsymbol{\nu}\, \mathrm{d}A$$
$$= \hat{e}^{\mathrm{i}}(s)\, \boldsymbol{\alpha} \cdot \hat{\boldsymbol{E}}^{\mathrm{T};\infty}(-\boldsymbol{\beta}, s)/s\mu_0 \tag{5.1}$$

Subsequently, the reciprocity theorem of the time-convolution type is applied to the unbounded domain exterior to the antenna system and to the scattered field and transmitted field states (see Table 5.1). Since both interrelated field states are causal, we may directly apply (1.44) and write

$$\int_{\boldsymbol{x}\in\mathcal{S}_0} \left(\hat{\boldsymbol{E}}^{\mathrm{s}} \times \hat{\boldsymbol{H}}^{\mathrm{T}} - \hat{\boldsymbol{E}}^{\mathrm{T}} \times \hat{\boldsymbol{H}}^{\mathrm{s}}\right) \cdot \boldsymbol{\nu}\, \mathrm{d}A = 0 \tag{5.2}$$

In view of Eq. (3.3), the sum of Eqs. (5.1) and (5.2) leads to

$$\int_{\boldsymbol{x}\in\mathcal{S}_0} \left(\hat{\boldsymbol{E}}^{\mathrm{R}} \times \hat{\boldsymbol{H}}^{\mathrm{T}} - \hat{\boldsymbol{E}}^{\mathrm{T}} \times \hat{\boldsymbol{H}}^{\mathrm{R}}\right) \cdot \boldsymbol{\nu}\, \mathrm{d}A$$
$$= \hat{e}^{\mathrm{i}}(s)\, \boldsymbol{\alpha} \cdot \hat{\boldsymbol{E}}^{\mathrm{T};\infty}(-\boldsymbol{\beta}, s)/s\mu_0 \tag{5.3}$$

Table 5.2 Application of the reciprocity theorem.

Domain $\mathcal{A}$		
Time-convolution	State (R)	State (T)
Source	0	0
Field	$\{\hat{\boldsymbol{E}}^{\mathrm{R}}, \hat{\boldsymbol{H}}^{\mathrm{R}}\}$	$\{\hat{\boldsymbol{E}}^{\mathrm{T}}, \hat{\boldsymbol{H}}^{\mathrm{T}}\}$
Material	$\{\hat{\underline{\eta}}, \hat{\underline{\zeta}}\}$	$\{\hat{\underline{\eta}}, \hat{\underline{\zeta}}\}$

In the final step, the reciprocity of the time-convolution type is applied to the bounded domain $\mathcal{A}$ in between $\mathcal{S}_0$ and $\mathcal{S}_1$ and to the total fields in the receiving and transmitting situations (see Table 5.2). In this way, we get:

$$\begin{aligned}
&\int_{\boldsymbol{x}\in\mathcal{S}_0} \left(\hat{\boldsymbol{E}}^{\mathrm{R}} \times \hat{\boldsymbol{H}}^{\mathrm{T}} - \hat{\boldsymbol{E}}^{\mathrm{T}} \times \hat{\boldsymbol{H}}^{\mathrm{R}}\right) \cdot \boldsymbol{\nu}\,\mathrm{d}A \\
&= \int_{\boldsymbol{x}\in\mathcal{S}_1} \left(\hat{\boldsymbol{E}}^{\mathrm{R}} \times \hat{\boldsymbol{H}}^{\mathrm{T}} - \hat{\boldsymbol{E}}^{\mathrm{T}} \times \hat{\boldsymbol{H}}^{\mathrm{R}}\right) \cdot \boldsymbol{\nu}\,\mathrm{d}A \\
&\quad + \int_{\boldsymbol{x}\in\mathcal{A}} \Big\{ \hat{\boldsymbol{H}}^{\mathrm{R}} \cdot \left[\hat{\underline{\zeta}} - \hat{\underline{\zeta}}^{T}\right] \cdot \hat{\boldsymbol{H}}^{\mathrm{T}} \\
&\qquad - \hat{\boldsymbol{E}}^{\mathrm{R}} \cdot \left[\hat{\underline{\eta}} - \hat{\underline{\eta}}^{T}\right] \cdot \hat{\boldsymbol{E}}^{\mathrm{T}} \Big\} \mathrm{d}V
\end{aligned} \tag{5.4}$$

where the volume integral vanishes for the medium that is self-adjoint in its EM properties (see Eqs. (1.31) and (1.32)). This includes antenna systems described with symmetric tensors of electric permittivity, conductivity and magnetic permeability, antenna systems made from PEC components (e.g., Figure 1.7), or, eventually, the combination of these cases. Since the surface integral over the terminal surface of the antenna system can be expressed in terms of Kirchhoff quantities, namely

$$\begin{aligned}
&\int_{\boldsymbol{x}\in\mathcal{S}_1} \left(\hat{\boldsymbol{E}}^{\mathrm{R}} \times \hat{\boldsymbol{H}}^{\mathrm{T}} - \hat{\boldsymbol{E}}^{\mathrm{T}} \times \hat{\boldsymbol{H}}^{\mathrm{R}}\right) \cdot \boldsymbol{\nu}\,\mathrm{d}A \\
&\qquad \simeq \sum_{m=1}^{N} \left[\hat{V}_m^{\mathrm{R}}(s)\hat{I}_m^{\mathrm{T}}(s) + \hat{V}_m^{\mathrm{T}}(s)\hat{I}_m^{\mathrm{R}}(s)\right]
\end{aligned} \tag{5.5}$$

we may, upon combining Eqs. (5.3)–(5.5), write

$$\sum_{m=1}^{N} \left[\hat{V}_m^{\mathrm{R}}(s)\hat{I}_m^{\mathrm{T}}(s) + \hat{V}_m^{\mathrm{T}}(s)\hat{I}_m^{\mathrm{R}}(s)\right] = \hat{e}^{\mathrm{i}}(s)\,\boldsymbol{\alpha} \cdot \hat{\boldsymbol{E}}^{\mathrm{T};\infty}(-\boldsymbol{\beta}, s)/s\mu_0 \tag{5.6}$$

for antenna systems with *reciprocal*[1] EM constitutive properties. Equation (5.6) is the starting point for constructing both Thévenin's and Norton's equivalent network representations of the receiving antenna system irradiated by the plane wave.

Thévenin's Network

Thévenin's network representation of the N-port receiving antenna system excited by the plane wave is arrived at with the help of the effective length of the current-excited transmitting antenna system

$$\hat{\boldsymbol{E}}^{\mathrm{T};\infty}(\boldsymbol{\xi}, s) = s\mu_0 \sum_{m=1}^{N} \hat{I}_m^{\mathrm{T}}(s)\hat{\boldsymbol{\ell}}_m^{\mathrm{T;I}}(\boldsymbol{\xi}, s) \quad \text{(4.23 revisited)}$$

$$\hat{\boldsymbol{H}}^{\mathrm{T};\infty}(\boldsymbol{\xi}, s) = sc_0^{-1} \sum_{m=1}^{N} \hat{I}_m^{\mathrm{T}}(s)\boldsymbol{\xi} \times \hat{\boldsymbol{\ell}}_m^{\mathrm{T;I}}(\boldsymbol{\xi}, s) \tag{5.7}$$

and Eq. (4.2) in the reciprocity relation (5.6). Upon rearranging the terms, we end up with

$$\sum_{n=1}^{N} \hat{Z}_{n,m}^{\mathrm{T}}(s)\hat{I}_n^{\mathrm{R}}(s) + \hat{V}_m^{\mathrm{R}}(s) = \hat{V}_m^{\mathrm{G}}(s) \tag{5.8}$$

for $m = \{1, \ldots, N\}$, where

$$\hat{V}_m^{\mathrm{G}}(s) = \hat{e}^{\mathrm{i}}(s)\,\boldsymbol{\alpha} \cdot \boldsymbol{\ell}_m^{\mathrm{T;I}}(-\boldsymbol{\beta}, s) \tag{5.9}$$

is the equivalent voltage-source strength.

Norton's Network

Similarly, Norton's network representation of the N-port receiving antenna system excited by the plane wave is arrived at with the help of the effective

1 The reciprocity relation (5.6) can also apply to nonreciprocal antenna systems provided that the appropriate change in the antenna constitutive parameters is made when switching from transmission to reception and vice versa [18].

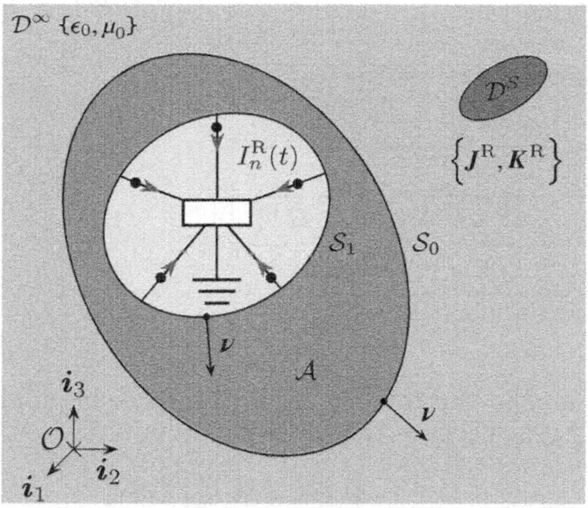

Figure 5.1 Receiving antenna system excited by a known volume-source distribution.

length of the voltage-excited transmitting antenna system:

$$\hat{\boldsymbol{H}}^{\mathrm{T};\infty}(\boldsymbol{\xi}, s) = s\epsilon_0 \sum_{m=1}^{N} \hat{V}_m^{\mathrm{T}}(s)\hat{\boldsymbol{e}}_m^{\mathrm{T;V}}(\boldsymbol{\xi}, s) \tag{5.10}$$

$$\hat{\boldsymbol{E}}^{\mathrm{T};\infty}(\boldsymbol{\xi}, s) = -sc_0^{-1} \sum_{m=1}^{N} \hat{V}_m^{\mathrm{T}}(s)\boldsymbol{\xi} \times \hat{\boldsymbol{e}}_m^{\mathrm{T;V}}(\boldsymbol{\xi}, s) \tag{5.11}$$

and Eq. (4.3) in the reciprocity relation (5.6). Upon rearranging the terms, we end up with

$$\sum_{n=1}^{N} \hat{Y}_{n,m}^{\mathrm{T}}(s)\hat{V}_n^{\mathrm{R}}(s) + \hat{I}_m^{\mathrm{R}}(s) = \hat{I}_m^{\mathrm{G}}(s) \tag{5.12}$$

for $m = \{1, \dots, N\}$, where

$$\hat{I}_m^{\mathrm{G}}(s) = \eta_0 \hat{e}^{\mathrm{i}}(s)\,(\boldsymbol{\alpha} \times \boldsymbol{\beta}) \cdot \boldsymbol{e}_m^{\mathrm{T;V}}(-\boldsymbol{\beta}, s) \tag{5.13}$$

is the equivalent current-source strength.

5.1.2 Equivalent Circuits for a Known Volume-Current Distribution

We shall analyze the receiving situation in which the antenna system is excited by defined electric and magnetic current volume densities whose (bounded) support $\mathcal{D}^S$ is located in the antenna embedding (see Figure 5.1). Applying the reciprocity theorem of the time-convolution type to the unbounded domain

Table 5.3 Application of the reciprocity theorem.

Domain $\mathcal{D}^\infty$		
Time-convolution	State (R)	State (T)
Source	$\{\hat{\boldsymbol{J}}^{\mathrm{R}}, \hat{\boldsymbol{K}}^{\mathrm{R}}\}$	0
Field	$\{\hat{\boldsymbol{E}}^{\mathrm{R}}, \hat{\boldsymbol{H}}^{\mathrm{R}}\}$	$\{\hat{\boldsymbol{E}}^{\mathrm{T}}, \hat{\boldsymbol{H}}^{\mathrm{T}}\}$
Material	$\{\epsilon_0, \mu_0\}$	$\{\epsilon_0, \mu_0\}$

exterior to the antenna system and to the total fields in the receiving and transmitting states according to Table 5.3, we arrive at

$$\int_{\boldsymbol{x}\in\mathcal{S}_0} \left(\hat{\boldsymbol{E}}^{\mathrm{R}} \times \hat{\boldsymbol{H}}^{\mathrm{T}} - \hat{\boldsymbol{E}}^{\mathrm{T}} \times \hat{\boldsymbol{H}}^{\mathrm{R}}\right) \cdot \boldsymbol{\nu}\,\mathrm{d}A$$

$$= \int_{\boldsymbol{x}\in\mathcal{D}^S} \left(\hat{\boldsymbol{K}}^{\mathrm{R}} \cdot \hat{\boldsymbol{H}}^{\mathrm{T}} - \hat{\boldsymbol{J}}^{\mathrm{R}} \cdot \hat{\boldsymbol{E}}^{\mathrm{T}}\right) \mathrm{d}V \tag{5.14}$$

where the integration over the outer bounding surface $\partial\mathcal{D}^\Delta$ vanishes as $\Delta \to \infty$ (see Section 1.4.3). Consequently, making use of Eqs. (5.4) and (5.5), we may write

$$\sum_{m=1}^{N} \left[\hat{V}_m^{\mathrm{R}}(s)\hat{I}_m^{\mathrm{T}}(s) + \hat{V}_m^{\mathrm{T}}(s)\hat{I}_m^{\mathrm{R}}(s)\right]$$

$$= \int_{\boldsymbol{x}\in\mathcal{D}^S} \left(\hat{\boldsymbol{K}}^{\mathrm{R}} \cdot \hat{\boldsymbol{H}}^{\mathrm{T}} - \hat{\boldsymbol{J}}^{\mathrm{R}} \cdot \hat{\boldsymbol{E}}^{\mathrm{T}}\right) \mathrm{d}V \tag{5.15}$$

for antenna systems with *reciprocal* EM constitutive properties. Equation (5.15) is the starting point for constructing both Thévenin's and Norton's equivalent network representations of the receiving antenna system excited by the defined volume-current distribution. The equivalent source strengths of these equivalent networks will be expressed via sensing EM wave fields represented by the impulse-excited radiated fields. These wave constituents are understood as configurational parameters that characterize the antenna's EM wave field transmission into its exterior domain.

Thévenin's Network

Thévenin's network representation of the N-port receiving antenna system excited by the known volume-source distribution is arrived at with the aid of the electric-current impulse-excited EM wave constituents defined as

$$\{\hat{\boldsymbol{E}}^{\mathrm{T}}, \hat{\boldsymbol{H}}^{\mathrm{T}}\}(\boldsymbol{x}, s) = \sum_{m=1}^{N} \hat{I}_m^{\mathrm{T}}(s)\{\hat{\boldsymbol{e}}_m^{\mathrm{T;I}}, \hat{\boldsymbol{h}}_m^{\mathrm{T;I}}\}(\boldsymbol{x}, s) \tag{5.16}$$

and Eq. (4.2) in the reciprocity relation (5.15). Upon rearranging the terms, we end up with (5.8) in which the equivalent voltage-source strength is given as

$$\hat{V}_m^{\mathrm{G}}(s) = \int_{\boldsymbol{x}\in\mathcal{D}^S} \left(\hat{\boldsymbol{K}}^{\mathrm{R}} \cdot \hat{\boldsymbol{h}}_m^{\mathrm{T;I}} - \hat{\boldsymbol{J}}^{\mathrm{R}} \cdot \hat{\boldsymbol{e}}_m^{\mathrm{T;I}}\right) \mathrm{d}V \tag{5.17}$$

for $m = \{1, \dots, N\}$.

Norton's Network

Norton's network representation of the N-port receiving antenna system excited by the known volume-source distribution is arrived at with the aid of the voltage impulse-excited EM wave constituents defined as

$$\{\hat{\boldsymbol{E}}^{\mathrm{T}}, \hat{\boldsymbol{H}}^{\mathrm{T}}\}(\boldsymbol{x}, s) = \sum_{m=1}^{N} \hat{V}_m^{\mathrm{T}}(s)\{\hat{\boldsymbol{e}}_m^{\mathrm{T;V}}, \hat{\boldsymbol{h}}_m^{\mathrm{T;V}}\}(\boldsymbol{x}, s) \tag{5.18}$$

and Eq. (4.3) in the reciprocity relation (5.15). Upon rearranging the terms, we end up with (5.12) in which the equivalent current-source strength is given as

$$\hat{I}_m^{\mathrm{G}}(s) = \int_{\boldsymbol{x}\in\mathcal{D}^S} \left(\hat{\boldsymbol{K}}^{\mathrm{R}} \cdot \hat{\boldsymbol{h}}_m^{\mathrm{T;V}} - \hat{\boldsymbol{J}}^{\mathrm{R}} \cdot \hat{\boldsymbol{e}}_m^{\mathrm{T;V}}\right) \mathrm{d}V \tag{5.19}$$

for $m = \{1, \dots, N\}$.

Exercise

- Give the relation between the absorption cross section of the load and the antenna gain concerning one and the same 1-port antenna system.

Hint: Define quantities

$$\hat{P}^{\mathrm{F}}(s) \triangleq \frac{1}{4}\left[\hat{V}^{\mathrm{T}}(s)\hat{I}^{\mathrm{T}}(-s) + \hat{V}^{\mathrm{T}}(-s)\hat{I}^{\mathrm{T}}(s)\right]$$
$$\hat{P}^{\mathrm{L}}(s) \triangleq \frac{1}{4}\left[\hat{V}^{\mathrm{R}}(s)\hat{I}^{\mathrm{R}}(-s) + \hat{V}^{\mathrm{R}}(-s)\hat{I}^{\mathrm{R}}(s)\right]$$

that for $\{s = \delta + \mathrm{i}\omega, \delta \downarrow 0, \omega \in \mathbb{R}\}$ has the meaning of the (time-averaged) power fed into the antenna and the power dissipated in the antenna load, respectively. The absorption cross section of the load is then defined as

$$\hat{\sigma}^{\mathrm{L}} \triangleq \hat{P}^{\mathrm{L}}/\hat{S}^{\mathrm{i}} \tag{5.20}$$

where $\hat{\boldsymbol{S}}^{\mathrm{i}}(s) = \boldsymbol{\beta}\hat{S}^{\mathrm{i}}(s) = \frac{1}{2}\boldsymbol{\beta}\,\eta_0\hat{e}^{\mathrm{i}}(s)\hat{e}^{\mathrm{i}}(-s)$ is the power-flow density of the incident plane wave. Notice that

$$\begin{aligned}
&16\hat{P}^{\mathrm{F}}(s)\hat{P}^{\mathrm{L}}(s)\\
&= \left[\hat{V}^{\mathrm{R}}(s)\hat{I}^{\mathrm{T}}(s) + \hat{V}^{\mathrm{T}}(s)\hat{I}^{\mathrm{R}}(s)\right]\left[\hat{V}^{\mathrm{R}}(-s)\hat{I}^{\mathrm{T}}(-s) + \hat{V}^{\mathrm{T}}(-s)\hat{I}^{\mathrm{R}}(-s)\right]\\
&- \left[\hat{V}^{\mathrm{R}}(s)\hat{I}^{\mathrm{T}}(-s) - \hat{V}^{\mathrm{T}}(-s)\hat{I}^{\mathrm{R}}(s)\right]\left[\hat{V}^{\mathrm{R}}(-s)\hat{I}^{\mathrm{T}}(s) - \hat{V}^{\mathrm{T}}(s)\hat{I}^{\mathrm{R}}(-s)\right]
\end{aligned}$$

which reveals the definition for efficiency of the load:

$$\begin{aligned}
\hat{\eta}^{\mathrm{L}}(s) &\triangleq \frac{16\hat{P}^{\mathrm{F}}(s)\hat{P}^{\mathrm{L}}(s)}{\left[\hat{V}^{\mathrm{R}}(s)\hat{I}^{\mathrm{T}}(s) + \hat{V}^{\mathrm{T}}(s)\hat{I}^{\mathrm{R}}(s)\right]\left[\hat{V}^{\mathrm{R}}(-s)\hat{I}^{\mathrm{T}}(-s) + \hat{V}^{\mathrm{T}}(-s)\hat{I}^{\mathrm{R}}(-s)\right]}\\
&= 1 - \frac{\left[\hat{Z}^{\mathrm{L}}(s) - \hat{Z}^{\mathrm{T}}(-s)\right]\left[\hat{Z}^{\mathrm{L}}(-s) - \hat{Z}^{\mathrm{T}}(s)\right]}{\left[\hat{Z}^{\mathrm{L}}(s) + \hat{Z}^{\mathrm{T}}(s)\right]\left[\hat{Z}^{\mathrm{L}}(-s) + \hat{Z}^{\mathrm{T}}(-s)\right]}
\end{aligned} \tag{5.21}$$

Obviously, $\{0 < \hat{\eta}^{\mathrm{L}} \leq 1\}$, where the equality applies to the "matched" load. Then use the reciprocity relation Eq. (5.6) for $N = 1$, namely,

$$\hat{V}^{\mathrm{R}}(s)\hat{I}^{\mathrm{T}}(s) + \hat{V}^{\mathrm{T}}(s)\hat{I}^{\mathrm{R}}(s) = \hat{e}^{\mathrm{i}}(s)\,\boldsymbol{\alpha} \cdot \hat{\boldsymbol{E}}^{\mathrm{T};\infty}(-\boldsymbol{\beta}, s)/s\mu_0 \tag{5.22}$$

and get

$$\frac{\hat{\sigma}^{\mathrm{L}}(-\boldsymbol{\beta}, s)}{\hat{\eta}^{\mathrm{L}}(s)} = -\frac{1}{s^2\mu_0^2}\eta_0^{-1}\frac{\boldsymbol{\alpha} \cdot \hat{\boldsymbol{E}}^{\mathrm{T};\infty}(-\boldsymbol{\beta}, s)\,\boldsymbol{\alpha} \cdot \hat{\boldsymbol{E}}^{\mathrm{T};\infty}(-\boldsymbol{\beta}, -s)}{8\hat{P}^{\mathrm{F}}(s)}$$

Consequently, define the transmitting antenna efficiency, namely,

$$\hat{\eta}^{\mathrm{T}}(s) \triangleq \hat{P}^{\mathrm{T}}(s)/\hat{P}^{\mathrm{F}}(s)$$

where $\hat{P}^{\mathrm{T}}(s)$ has the meaning of the radiated power that can be expressed in terms of the radiated far-field amplitude as

$$\hat{P}^{\mathrm{T}}(s) = \frac{\eta_0}{32\pi^2} \int_{\xi\in\Omega} \hat{\boldsymbol{E}}^{\mathrm{T};\infty}(\boldsymbol{\xi}, s) \cdot \hat{\boldsymbol{E}}^{\mathrm{T};\infty}(\boldsymbol{\xi}, -s)\mathrm{d}\Omega$$

Then write

$$\frac{\hat{\sigma}^{\mathrm{L}}(-\boldsymbol{\beta}, s)}{\hat{\eta}^{\mathrm{L}}(s)} = -\pi\frac{c_0^2}{s^2} \frac{\boldsymbol{\alpha} \cdot \hat{\boldsymbol{E}}^{\mathrm{T};\infty}(-\boldsymbol{\beta}, s)\, \boldsymbol{\alpha} \cdot \hat{\boldsymbol{E}}^{\mathrm{T};\infty}(-\boldsymbol{\beta}, -s)}{\int_{\xi\in\Omega} \hat{\boldsymbol{E}}^{\mathrm{T};\infty}(\boldsymbol{\xi}, s) \cdot \hat{\boldsymbol{E}}^{\mathrm{T};\infty}(\boldsymbol{\xi}, -s)\mathrm{d}\Omega} 4\pi\hat{\eta}^{\mathrm{T}}(s)$$

Multiplication of the latter by factor

$$\frac{\hat{\boldsymbol{E}}^{\mathrm{T};\infty}(-\boldsymbol{\beta}, s) \cdot \hat{\boldsymbol{E}}^{\mathrm{T};\infty}(-\boldsymbol{\beta}, -s)}{\hat{\boldsymbol{E}}^{\mathrm{T};\infty}(-\boldsymbol{\beta}, s) \cdot \hat{\boldsymbol{E}}^{\mathrm{T};\infty}(-\boldsymbol{\beta}, -s)}$$

suggests to define the antenna (power) gain

$$\hat{G}(\boldsymbol{\gamma}, s) \triangleq \hat{\eta}^{\mathrm{T}}(s)\hat{D}(\boldsymbol{\gamma}, s)$$

with the antenna directivity

$$\hat{D}(\boldsymbol{\gamma}, s) \triangleq \frac{4\pi\, \hat{\boldsymbol{E}}^{\mathrm{T};\infty}(\boldsymbol{\gamma}, s) \cdot \hat{\boldsymbol{E}}^{\mathrm{T};\infty}(\boldsymbol{\gamma}, -s)}{\int_{\xi\in\Omega} \hat{\boldsymbol{E}}^{\mathrm{T};\infty}(\boldsymbol{\xi}, s) \cdot \hat{\boldsymbol{E}}^{\mathrm{T};\infty}(\boldsymbol{\xi}, -s)\mathrm{d}\Omega}$$

and the polarization efficiency, namely,

$$\hat{\eta}^{\mathrm{P}}(\boldsymbol{\gamma}, s) \triangleq \frac{\boldsymbol{\alpha} \cdot \hat{\boldsymbol{E}}^{\mathrm{T};\infty}(\boldsymbol{\gamma}, s)\, \boldsymbol{\alpha} \cdot \hat{\boldsymbol{E}}^{\mathrm{T};\infty}(\boldsymbol{\gamma}, -s)}{\hat{\boldsymbol{E}}^{\mathrm{T};\infty}(\boldsymbol{\gamma}, s) \cdot \hat{\boldsymbol{E}}^{\mathrm{T};\infty}(\boldsymbol{\gamma}, -s)}$$

that attains its maximum, $\hat{\eta}^{\mathrm{P}} = 1$, when the polarization of the incident plane wave is "matched" to the state of polarization of the radiation characteristics. Consequently, we may get the desired "power–reciprocity relation":

$$\hat{\sigma}^{\mathrm{L}}(-\boldsymbol{\beta}, s) = -\pi c_0^2 s^{-2}\hat{\eta}^{\mathrm{L}}(s)\hat{\eta}^{\mathrm{P}}(-\boldsymbol{\beta}, s)\hat{G}(-\boldsymbol{\beta}, s)$$

that relates the absorption cross section of the load in the receiving state to the antenna gain in the corresponding transmitting situation. Thanks to the property

$$(4\pi)^{-1} \int_{\xi\in\Omega} \hat{D}(\boldsymbol{\xi}, s)\mathrm{d}\Omega = 1$$

and $\{0 \leq \hat{\eta}^{\mathrm{P}} \leq 1\}$, it is interesting to note that for the *average* absorption cross section of the load the following inequality applies:

$$(4\pi)^{-1} \int_{\xi \in \Omega} \hat{\sigma}^{\mathrm{L}}(\xi, s)\mathrm{d}\Omega \leq -\pi c_0^2 s^{-2} \hat{\eta}^{\mathrm{L}}(s)\hat{\eta}^{\mathrm{T}}(s)$$

Accordingly, for a loss-free antenna whose load and polarization are perfectly "matched," the average cross section of the load equals to $-\pi c_0^2 s^{-2}$, which for $s = \mathrm{i}\omega$ corresponds to $\lambda_0^2/4\pi$, where λ_0 is the corresponding wavelength. For more details on the subject, we refer the reader to Ref. [18].

6

The Antenna System in the Presence of a Scatterer

In real-life scenarios, antenna systems do operate in the vicinity of objects that may electromagnetically couple to these systems and thus change their EM characteristics. From this reason, it is of high practical importance to have a description of the impact of an object on the antenna behavior that will make possible to quantitatively assess potential consequences with regard to EM interference and EM susceptibility aspects. This is exactly the main objective of this chapter, where the relevant reciprocity analysis is carried out.

6.1 Receiving Antenna in the Presence of a Scatterer

In this section, we shall analyze two receiving situations differing from each other in the presence of a scatterer that is located in the antenna embedding (see Figure 6.1). The scatterer occupies domain D whose bounding surface is denoted by ∂D. The receiving antenna system is in both receiving scenarios, say (R) and ($\tilde{\mathrm{R}}$), excited either by the uniform plane wave

$$\hat{\boldsymbol{E}}^{\mathrm{i}}(\boldsymbol{x}, s) = \boldsymbol{\alpha}\, \hat{e}^{\mathrm{i}}(s) \exp(-s\boldsymbol{\beta} \cdot \boldsymbol{x}/c_0) \qquad \text{(3.1 revisited)}$$

$$\hat{\boldsymbol{H}}^{\mathrm{i}}(\boldsymbol{x}, s) = (\boldsymbol{\beta} \times \boldsymbol{\alpha})\, \eta_0 \hat{e}^{\mathrm{i}}(s) \exp(-s\boldsymbol{\beta} \cdot \boldsymbol{x}/c_0) \qquad \text{(3.2 revisited)}$$

or by a defined volume source distribution generating the incident EM wave field $\{\hat{\boldsymbol{E}}^{\mathrm{i}}, \hat{\boldsymbol{H}}^{\mathrm{i}}\}(\boldsymbol{x}, s)$. As a consequence of the scatterer in scenario ($\tilde{\mathrm{R}}$), the scattered EM wave fields in the configurations are different and we may hence define their difference, namely

$$\{\Delta\hat{\boldsymbol{E}}^{\mathrm{s}}, \Delta\hat{\boldsymbol{H}}^{\mathrm{s}}\}(\boldsymbol{x}, s) \triangleq \{\hat{\boldsymbol{E}}^{\tilde{\mathrm{s}}} - \hat{\boldsymbol{E}}^{\mathrm{s}}, \hat{\boldsymbol{H}}^{\tilde{\mathrm{s}}} - \hat{\boldsymbol{H}}^{\mathrm{s}}\}(\boldsymbol{x}, s) \qquad (6.1)$$

Since the incident EM wave field remains the same, the change in the scattered field equals to the change in the total field, that is,

$$\begin{aligned}\{\Delta\hat{\boldsymbol{E}}^{\mathrm{R}}, \Delta\hat{\boldsymbol{H}}^{\mathrm{R}}\}(\boldsymbol{x}, s) &\triangleq \{\hat{\boldsymbol{E}}^{\tilde{\mathrm{R}}} - \hat{\boldsymbol{E}}^{\mathrm{R}}, \hat{\boldsymbol{H}}^{\tilde{\mathrm{R}}} - \hat{\boldsymbol{H}}^{\mathrm{R}}\}(\boldsymbol{x}, s)\\ &= \{\Delta\hat{\boldsymbol{E}}^{\mathrm{s}}, \Delta\hat{\boldsymbol{H}}^{\mathrm{s}}\}(\boldsymbol{x}, s) \qquad (6.2)\end{aligned}$$

Electromagnetic Reciprocity in Antenna Theory, First Edition. Martin Stumpf.

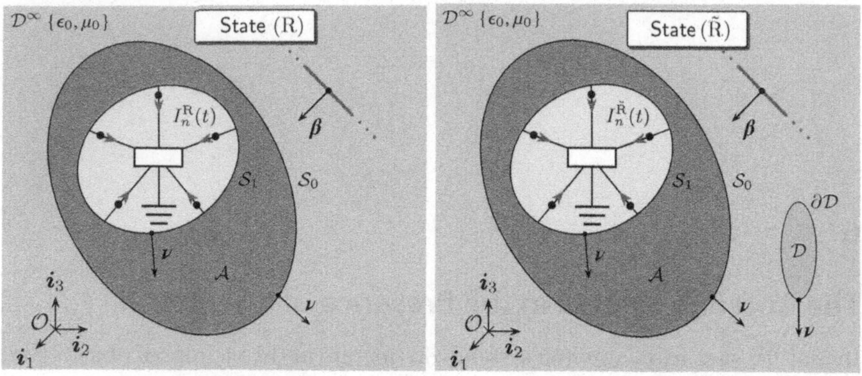

Figure 6.1 Receiving situations differing from each other in the presence of a scatterer.

The corresponding EM field states will be denoted by (Δs) and (ΔR). In order to describe the resulting change in the strength of internal generators of the Kirchhoff-type equivalent circuits, the receiving states shown in Figure 6.1 will be interrelated with the corresponding transmitting situation without the scatterer (see state (T) in Figure 4.1). Accordingly, we start with applying the reciprocity theorem of the time-convolution type to the unbounded domain exterior to the antenna system and the scatterer and to the field-difference state (ΔR) and the transmitting state (T) (see Table 6.1). Since both fields are causal and source-free outside $\mathcal{S}_0 \cup \partial\mathcal{D}$, we can make use of the conclusions drawn in Section 1.4.3 and get

$$\int_{\boldsymbol{x}\in\mathcal{S}_0\cup\partial\mathcal{D}} \left(\Delta\hat{\boldsymbol{E}}^{\mathrm{R}} \times \hat{\boldsymbol{H}}^{\mathrm{T}} - \hat{\boldsymbol{E}}^{\mathrm{T}} \times \Delta\hat{\boldsymbol{H}}^{\mathrm{R}}\right) \cdot \boldsymbol{\nu}\,\mathrm{d}A = 0 \tag{6.3}$$

In the following step, the reciprocity of the time-convolution type is applied to the total fields in states (R) and (T) and to the domain occupied by the scatterer $\mathcal{D}$ (see Table 6.2). Since these field states are source-free here and since the embedding is self-adjoint in its EM properties, we will end up with the zero contribution from the relevant surface integral over the bounding surface of the scatterer, that is,

$$\int_{\boldsymbol{x}\in\partial\mathcal{D}} \left(\hat{\boldsymbol{E}}^{\mathrm{R}} \times \hat{\boldsymbol{H}}^{\mathrm{T}} - \hat{\boldsymbol{E}}^{\mathrm{T}} \times \hat{\boldsymbol{H}}^{\mathrm{R}}\right) \cdot \boldsymbol{\nu}\,\mathrm{d}A = 0 \tag{6.4}$$

Table 6.1 Application of the reciprocity theorem.

Domain exterior to $\mathcal{S}_0 \cup \partial\mathcal{D}$		
Time-convolution	State $(\Delta\mathrm{R})$	State (T)
Source	0	0
Field	$\{\Delta\hat{\boldsymbol{E}}^{\mathrm{R}}, \Delta\hat{\boldsymbol{H}}^{\mathrm{R}}\}$	$\{\hat{\boldsymbol{E}}^{\mathrm{T}}, \hat{\boldsymbol{H}}^{\mathrm{T}}\}$
Material	$\{\epsilon_0, \mu_0\}$	$\{\epsilon_0, \mu_0\}$

Combining Eqs. (6.3) and (6.4) with Eq. (6.2), we obtain

$$\begin{aligned}\int_{\boldsymbol{x}\in\mathcal{S}_0} &\left(\hat{\boldsymbol{E}}^{\tilde{\mathrm{R}}} \times \hat{\boldsymbol{H}}^{\mathrm{T}} - \hat{\boldsymbol{E}}^{\mathrm{T}} \times \hat{\boldsymbol{H}}^{\tilde{\mathrm{R}}}\right) \cdot \boldsymbol{\nu}\,\mathrm{d}A \\ &- \int_{\boldsymbol{x}\in\mathcal{S}_0} \left(\hat{\boldsymbol{E}}^{\mathrm{R}} \times \hat{\boldsymbol{H}}^{\mathrm{T}} - \hat{\boldsymbol{E}}^{\mathrm{T}} \times \hat{\boldsymbol{H}}^{\mathrm{R}}\right) \cdot \boldsymbol{\nu}\,\mathrm{d}A \\ &\qquad = -\int_{\boldsymbol{x}\in\partial\mathcal{D}} \left(\hat{\boldsymbol{E}}^{\tilde{\mathrm{R}}} \times \hat{\boldsymbol{H}}^{\mathrm{T}} - \hat{\boldsymbol{E}}^{\mathrm{T}} \times \hat{\boldsymbol{H}}^{\tilde{\mathrm{R}}}\right) \cdot \boldsymbol{\nu}\,\mathrm{d}A \end{aligned} \tag{6.5}$$

Next, following the procedure described in Section 5.1, we may rewrite the latter as

$$\begin{aligned}\sum_{m=1}^{N} &\left[\hat{V}_m^{\tilde{\mathrm{R}}}(s)\hat{I}_m^{\mathrm{T}}(s) + \hat{V}_m^{\mathrm{T}}(s)\hat{I}_m^{\tilde{\mathrm{R}}}(s)\right] \\ &- \sum_{m=1}^{N} \left[\hat{V}_m^{\mathrm{R}}(s)\hat{I}_m^{\mathrm{T}}(s) + \hat{V}_m^{\mathrm{T}}(s)\hat{I}_m^{\mathrm{R}}(s)\right] \\ &\qquad = -\int_{\boldsymbol{x}\in\partial\mathcal{D}} \left(\hat{\boldsymbol{E}}^{\tilde{\mathrm{R}}} \times \hat{\boldsymbol{H}}^{\mathrm{T}} - \hat{\boldsymbol{E}}^{\mathrm{T}} \times \hat{\boldsymbol{H}}^{\tilde{\mathrm{R}}}\right) \cdot \boldsymbol{\nu}\,\mathrm{d}A \end{aligned} \tag{6.6}$$

Table 6.2 Application of the reciprocity theorem.

Domain $\mathcal{D}$		
Time-convolution	State (R)	State (T)
Source	0	0
Field	$\{\hat{\boldsymbol{E}}^{\mathrm{R}}, \hat{\boldsymbol{H}}^{\mathrm{R}}\}$	$\{\hat{\boldsymbol{E}}^{\mathrm{T}}, \hat{\boldsymbol{H}}^{\mathrm{T}}\}$
Material	$\{\epsilon_0, \mu_0\}$	$\{\epsilon_0, \mu_0\}$

Table 6.3 Application of the reciprocity theorem.

Domain $\mathcal{D}$		
Time-convolution	State ($\tilde{\mathrm{R}}$)	State (T)
Source	$\{\hat{\boldsymbol{J}}^{\tilde{\mathrm{R}}}, \hat{\boldsymbol{K}}^{\tilde{\mathrm{R}}}\}$	0
Field	$\{\hat{\boldsymbol{E}}^{\tilde{\mathrm{R}}}, \hat{\boldsymbol{H}}^{\tilde{\mathrm{R}}}\}$	$\{\hat{\boldsymbol{E}}^{\mathrm{T}}, \hat{\boldsymbol{H}}^{\mathrm{T}}\}$
Material	$\{\epsilon_0, \mu_0\}$	$\{\epsilon_0, \mu_0\}$

for antenna systems that are self-adjoint in the EM constitutive parameters. The right-hand side of the latter can be further specified. For EM-penetrable scatterers whose EM properties are described by the (tensorial) transverse admittance and the longitudinal impedance, $\underline{\hat{\boldsymbol{\eta}}}^{\mathrm{s}} = \underline{\hat{\boldsymbol{\eta}}}^{\mathrm{s}}(\boldsymbol{x}, s)$ and $\underline{\hat{\boldsymbol{\zeta}}}^{\mathrm{s}} = \underline{\hat{\boldsymbol{\zeta}}}^{\mathrm{s}}(\boldsymbol{x}, s)$ (see Eqs. (1.23) and (1.24)), respectively, the application of the reciprocity theorem of the time-convolution type according to Table 6.3 leads to

$$\int_{\boldsymbol{x}\in\partial\mathcal{D}} \left(\hat{\boldsymbol{E}}^{\tilde{\mathrm{R}}} \times \hat{\boldsymbol{H}}^{\mathrm{T}} - \hat{\boldsymbol{E}}^{\mathrm{T}} \times \hat{\boldsymbol{H}}^{\tilde{\mathrm{R}}}\right) \cdot \boldsymbol{\nu}\,\mathrm{d}A$$
$$= \int_{\boldsymbol{x}\in\mathcal{D}} \Big\{\hat{\boldsymbol{E}}^{\mathrm{T}} \cdot \left[\underline{\hat{\boldsymbol{\eta}}}^{\mathrm{s}} - s\epsilon_0\underline{\boldsymbol{I}}\right] \cdot \hat{\boldsymbol{E}}^{\tilde{\mathrm{R}}}$$
$$- \hat{\boldsymbol{H}}^{\mathrm{T}} \cdot \left[\underline{\hat{\boldsymbol{\zeta}}}^{\mathrm{s}} - s\mu_0\underline{\boldsymbol{I}}\right] \cdot \hat{\boldsymbol{H}}^{\tilde{\mathrm{R}}}\Big\}\mathrm{d}V \tag{6.7}$$

where the presence of the scatterer has been accounted for via the equivalent contrast electric and magnetic current volume densities, specifically

$$\hat{\boldsymbol{J}}^{\tilde{\mathrm{R}}}(\boldsymbol{x}, s) = [\,\underline{\hat{\boldsymbol{\eta}}}^{\mathrm{s}}(\boldsymbol{x}, s) - s\epsilon_0\underline{\boldsymbol{I}}\,] \cdot \hat{\boldsymbol{E}}^{\tilde{\mathrm{R}}}(\boldsymbol{x}, s) \tag{6.8}$$

$$\hat{\boldsymbol{K}}^{\tilde{\mathrm{R}}}(\boldsymbol{x}, s) = [\,\underline{\hat{\boldsymbol{\zeta}}}^{\mathrm{s}}(\boldsymbol{x}, s) - s\mu_0\underline{\boldsymbol{I}}\,] \cdot \hat{\boldsymbol{H}}^{\tilde{\mathrm{R}}}(\boldsymbol{x}, s) \tag{6.9}$$

for $\boldsymbol{x} \in \mathcal{D}$. Finally, for PEC and PMC EM impenetrable scatterers, the following explicit-type boundary conditions apply:

$$\lim_{\delta\downarrow 0} \boldsymbol{\nu} \times \hat{\boldsymbol{E}}^{\tilde{\mathrm{R}}}(\boldsymbol{x} + \delta\boldsymbol{\nu}, s) = \boldsymbol{0} \tag{6.10}$$

$$\lim_{\delta\downarrow 0} \boldsymbol{\nu} \times \hat{\boldsymbol{H}}^{\tilde{\mathrm{R}}}(\boldsymbol{x} + \delta\boldsymbol{\nu}, s) = \boldsymbol{0} \tag{6.11}$$

for $\boldsymbol{x} \in \partial\mathcal{D}$, respectively. Equations (6.6)–(6.11) will next serve as the starting point for expressing the change of source strength of the Kirchhoff-type equivalent circuits.

Thévenin's Network

The change in the voltage-source strengths of the equivalent Thévenin circuit generator is defined as the difference between the open-circuit voltages in the configuration with and without the scatterer, that is,

$$\Delta\hat{V}_m^{\mathrm{G}}(s) \triangleq \hat{V}_m^{\tilde{\mathrm{G}}}(s) - \hat{V}_m^{\mathrm{G}}(s) \tag{6.12}$$

for all $m = \{1, \dots, N\}$. Consequently, making use of Eq. (5.8) in Eq. (6.6) along with Eq. (5.16), the change in the voltage-source strength follows:

$$\Delta\hat{V}_m^{\mathrm{G}}(s) = \int_{\boldsymbol{x}\in\partial\mathcal{D}} \left(\hat{\boldsymbol{e}}_m^{\mathrm{T;I}} \times \hat{\boldsymbol{H}}^{\tilde{\mathrm{R}}} - \hat{\boldsymbol{E}}^{\tilde{\mathrm{R}}} \times \hat{\boldsymbol{h}}_m^{\mathrm{T;I}}\right) \cdot \boldsymbol{\nu}\,\mathrm{d}A \tag{6.13}$$

for $m = \{1, \dots, N\}$. In accordance with Eq. (6.7), the latter expression has its special form for EM penetrable scatterers, namely,

$$\Delta\hat{V}_m^{\mathrm{G}}(s) = -\int_{\boldsymbol{x}\in\mathcal{D}} \left\{\hat{\boldsymbol{e}}_m^{\mathrm{T;I}} \cdot \left[\underline{\hat{\boldsymbol{\eta}}}^{\mathrm{s}} - s\epsilon_0\underline{\boldsymbol{I}}\right] \cdot \hat{\boldsymbol{E}}^{\tilde{\mathrm{R}}} - \hat{\boldsymbol{h}}_m^{\mathrm{T;I}} \cdot \left[\underline{\hat{\boldsymbol{\zeta}}}^{\mathrm{s}} - s\mu_0\underline{\boldsymbol{I}}\right] \cdot \hat{\boldsymbol{H}}^{\tilde{\mathrm{R}}}\right\}\mathrm{d}V \tag{6.14}$$

for $m = \{1, \dots, N\}$. Also, making use of the boundary conditions (6.10) and (6.11), one may express the change as

$$\Delta\hat{V}_m^{\mathrm{G}}(s) = -\int_{\boldsymbol{x}\in\partial\mathcal{D}} \hat{\boldsymbol{e}}_m^{\mathrm{T;I}} \cdot \partial\hat{\boldsymbol{J}}^{\tilde{\mathrm{R}}}\,\mathrm{d}A \tag{6.15}$$

for a PEC scatterer with $m = \{1, \dots, N\}$ and as

$$\Delta\hat{V}_m^{\mathrm{G}}(s) = \int_{\boldsymbol{x}\in\partial\mathcal{D}} \hat{\boldsymbol{h}}_m^{\mathrm{T;I}} \cdot \partial\hat{\boldsymbol{K}}^{\tilde{\mathrm{R}}}\,\mathrm{d}A \tag{6.16}$$

for a PMC scatterer with $m = \{1, \dots, N\}$, where we have defined the equivalent electric and magnetic current surface densities:

$$\partial\hat{\boldsymbol{J}}^{\tilde{\mathrm{R}}}(\boldsymbol{x}, s) \triangleq \boldsymbol{\nu}(\boldsymbol{x}) \times \hat{\boldsymbol{H}}^{\tilde{\mathrm{R}}}(\boldsymbol{x}, s) \tag{6.17}$$

$$\partial\hat{\boldsymbol{K}}^{\tilde{\mathrm{R}}}(\boldsymbol{x}, s) \triangleq \hat{\boldsymbol{E}}^{\tilde{\mathrm{R}}}(\boldsymbol{x}, s) \times \boldsymbol{\nu}(\boldsymbol{x}) \tag{6.18}$$

where surface $\partial\mathcal{D}$ is approached from its exterior. A special form of Eq. (6.14) under the first-order Rayleigh–Gans–Born approximation ([16], Sec. 29.5) has been previously applied to evaluate the effect of EM scattering from raindrops to EM wave propagation ([9], Sec. 6.8).

Norton's Network

Similar to the previous section, we start with the definition of the change in the electric current-source strength of the Norton circuit generators

$$\Delta\hat{I}_m^{\mathrm{G}}(s) \triangleq \hat{I}_m^{\tilde{\mathrm{G}}}(s) - \hat{I}_m^{\mathrm{G}}(s) \tag{6.19}$$

for $m = \{1, \ldots, N\}$, where $\hat{I}_m^{\tilde{\mathrm{G}}}$ and $\hat{I}_m^{\mathrm{G}}$ are short-circuit electric currents that apply to the case with and without the scatterer, respectively. Consequently, making use of Eq. (5.12) in Eq. (6.6) along with Eq. (5.18), the change in the current-source strength follows:

$$\Delta\hat{I}_m^{\mathrm{G}}(s) = \int_{\boldsymbol{x}\in\partial D} \left(\hat{\boldsymbol{e}}_m^{\mathrm{T;V}} \times \hat{\boldsymbol{H}}^{\tilde{\mathrm{R}}} - \hat{\boldsymbol{E}}^{\tilde{\mathrm{R}}} \times \hat{\boldsymbol{h}}_m^{\mathrm{T;V}}\right) \cdot \boldsymbol{\nu}\, \mathrm{d}A \tag{6.20}$$

for $m = \{1, \ldots, N\}$. In accordance with Eq. (6.7), the latter expression has its special form for EM penetrable scatterers, namely,

$$\Delta\hat{I}_m^{\mathrm{G}}(s) = -\int_{\boldsymbol{x}\in D} \left\{\hat{\boldsymbol{e}}_m^{\mathrm{T;V}} \cdot \left[\underline{\hat{\boldsymbol{\eta}}}^{\mathrm{s}} - s\epsilon_0 \underline{\boldsymbol{I}}\right] \cdot \hat{\boldsymbol{E}}^{\tilde{\mathrm{R}}} - \hat{\boldsymbol{h}}_m^{\mathrm{T;V}} \cdot \left[\underline{\hat{\boldsymbol{\zeta}}}^{\mathrm{s}} - s\mu_0 \underline{\boldsymbol{I}}\right] \cdot \hat{\boldsymbol{H}}^{\tilde{\mathrm{R}}}\right\} \mathrm{d}V \tag{6.21}$$

Finally, making use of the boundary conditions (6.10) and (6.11), one may express the change as

$$\Delta\hat{I}_m^{\mathrm{G}}(s) = -\int_{\boldsymbol{x}\in\partial D} \hat{\boldsymbol{e}}_m^{\mathrm{T;V}} \cdot \partial\hat{\boldsymbol{J}}^{\tilde{\mathrm{R}}}\, \mathrm{d}A \tag{6.22}$$

for a PEC scatterer with $m = \{1, \ldots, N\}$ and as

$$\Delta\hat{I}_m^{\mathrm{G}}(s) = \int_{\boldsymbol{x}\in\partial D} \hat{\boldsymbol{h}}_m^{\mathrm{T;V}} \cdot \partial\hat{\boldsymbol{K}}^{\tilde{\mathrm{R}}}\, \mathrm{d}A \tag{6.23}$$

for a PMC scatterer with $m = \{1, \ldots, N\}$.

6.2 Transmitting Antenna in the Presence of a Scatterer

In this section, we shall describe the impact of a scatterer on the matrix of the antenna radiation impedance (see Eq. (4.2)). To this end, we will mutually interrelate two transmitting states depicted in Figure 6.2. Here, two N-port transmitting antenna systems are excited at their accessible ports by either electric currents $I_n^{\mathrm{T}}(t)$ or voltages $V_n^{\mathrm{T}}(t)$. The way of excitation is identical in both scenarios. Thanks to the presence of the scatterer whose EM properties differ from those of the embedding, the total radiated fields in the transmitting situations are different. This difference is expressed as

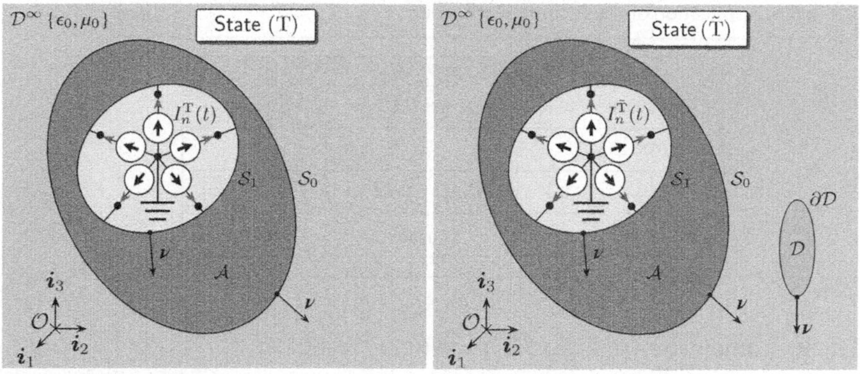

Figure 6.2 Transmitting situations differing from each other in the presence of a scatterer.

$$\{\Delta\hat{\boldsymbol{E}}^{\mathrm{T}}, \Delta\hat{\boldsymbol{H}}^{\mathrm{T}}\}(\boldsymbol{x}, s) \triangleq \{\hat{\boldsymbol{E}}^{\tilde{\mathrm{T}}} - \hat{\boldsymbol{E}}^{\mathrm{T}}, \hat{\boldsymbol{H}}^{\tilde{\mathrm{T}}} - \hat{\boldsymbol{H}}^{\mathrm{T}}\}(\boldsymbol{x}, s) \tag{6.24}$$

for $\boldsymbol{x} \in \mathcal{A} \cup \mathcal{D}^{\infty}$. The corresponding EM wave field states are denoted by (T), ($\tilde{\mathrm{T}}$) and (ΔT).

6.2.1 Analysis Based on the Reciprocity Theorem of the Time-Convolution Type

The reciprocity analysis starts by applying the reciprocity theorem of the time-convolution type to the unbounded domain exterior to the antenna system and the scatterer and to the field-difference state (ΔT) and the transmitting state (T) in absence of the scatterer (see Table 6.4). In this way, making use of the conclusions drawn in Section 1.4.3, we end up with

$$\int_{\boldsymbol{x}\in\mathcal{S}_0\cup\partial\mathcal{D}} \left(\Delta\hat{\boldsymbol{E}}^{\mathrm{T}} \times \hat{\boldsymbol{H}}^{\mathrm{T}} - \hat{\boldsymbol{E}}^{\mathrm{T}} \times \Delta\hat{\boldsymbol{H}}^{\mathrm{T}}\right) \cdot \boldsymbol{\nu}\,\mathrm{d}A = 0 \tag{6.25}$$

The application of the reciprocity theorem of the time-convolution type to the domain occupied by the antenna system leads to

$$\begin{aligned}\int_{\boldsymbol{x}\in\mathcal{S}_0} &\left(\Delta\hat{\boldsymbol{E}}^{\mathrm{T}} \times \hat{\boldsymbol{H}}^{\mathrm{T}} - \hat{\boldsymbol{E}}^{\mathrm{T}} \times \Delta\hat{\boldsymbol{H}}^{\mathrm{T}}\right) \cdot \boldsymbol{\nu}\,\mathrm{d}A \\ &= \int_{\boldsymbol{x}\in\mathcal{S}_1} \left(\Delta\hat{\boldsymbol{E}}^{\mathrm{T}} \times \hat{\boldsymbol{H}}^{\mathrm{T}} - \hat{\boldsymbol{E}}^{\mathrm{T}} \times \Delta\hat{\boldsymbol{H}}^{\mathrm{T}}\right) \cdot \boldsymbol{\nu}\,\mathrm{d}A \\ &\simeq \sum_{m=1}^{N} \left[\Delta\hat{V}_m^{\mathrm{T}}(s)\hat{I}_m^{\mathrm{T}}(s) - \hat{V}_m^{\mathrm{T}}(s)\Delta\hat{I}_m^{\mathrm{T}}(s)\right]\end{aligned} \tag{6.26}$$

for antenna systems with self-adjoint EM properties (see Section 5.1.1).

Table 6.4 Application of the reciprocity theorem.

Domain exterior to $\mathcal{S}_0 \cup \partial\mathcal{D}$		
Time-convolution	state (ΔT)	state (T)
Source	0	0
Field	$\{\Delta\hat{\boldsymbol{E}}^{\mathrm{T}}, \Delta\hat{\boldsymbol{H}}^{\mathrm{T}}\}$	$\{\hat{\boldsymbol{E}}^{\mathrm{T}}, \hat{\boldsymbol{H}}^{\mathrm{T}}\}$
Material	$\{\epsilon_0, \mu_0\}$	$\{\epsilon_0, \mu_0\}$

Finally, combination of Eqs. (6.25) and (6.26) results in

$$\sum_{m=1}^{N} \left[\Delta\hat{V}_m^{\mathrm{T}}(s)\hat{I}_m^{\mathrm{T}}(s) - \hat{V}_m^{\mathrm{T}}(s)\Delta\hat{I}_m^{\mathrm{T}}(s)\right]$$
$$= \int_{\boldsymbol{x}\in\partial\mathcal{D}} \left(\hat{\boldsymbol{E}}^{\mathrm{T}} \times \hat{\boldsymbol{H}}^{\tilde{\mathrm{T}}} - \hat{\boldsymbol{E}}^{\tilde{\mathrm{T}}} \times \hat{\boldsymbol{H}}^{\mathrm{T}}\right) \cdot \boldsymbol{\nu}\,\mathrm{d}A \qquad (6.27)$$

The latter relation is the starting point for describing the change in the antenna radiation impedance and admittance matrices.

Thévenin's Network

The change in the radiation impedance matrix of the Thévenin network representation is in line with Section 5.1.2 described using the electric current-excited sensing-field EM wave field constituents. To this end, we take $\Delta\hat{I}_m^{\mathrm{T}} = 0$ for all $m = \{1, \ldots, N\}$ and use Eq. (4.2) in the left-hand side of Eq. (6.27) and get

$$\sum_{m=1}^{N}\sum_{n=1}^{N} \hat{I}_m^{\mathrm{T}}(s)\Delta\hat{Z}_{m,n}^{\mathrm{T}}(s)\hat{I}_n^{\mathrm{T}}(s)$$
$$= \sum_{m=1}^{N}\sum_{n=1}^{N} \hat{I}_m^{\mathrm{T}}(s)\hat{I}_n^{\mathrm{T}}(s) \int_{\boldsymbol{x}\in\partial\mathcal{D}} \left(\hat{\boldsymbol{e}}_m^{\mathrm{T;I}} \times \hat{\boldsymbol{h}}_n^{\tilde{\mathrm{T}};\mathrm{I}} - \hat{\boldsymbol{e}}_n^{\tilde{\mathrm{T}};\mathrm{I}} \times \hat{\boldsymbol{h}}_m^{\mathrm{T;I}}\right) \cdot \boldsymbol{\nu}\,\mathrm{d}A \qquad (6.28)$$

where we have used Eq. (5.16). Since the latter equality has to hold for arbitrary exciting electric currents, we finally end up with

$$\Delta\hat{Z}_{m,n}^{\mathrm{T}}(s) = \int_{\boldsymbol{x}\in\partial\mathcal{D}} \left(\hat{\boldsymbol{e}}_m^{\mathrm{T;I}} \times \hat{\boldsymbol{h}}_n^{\tilde{\mathrm{T}};\mathrm{I}} - \hat{\boldsymbol{e}}_n^{\tilde{\mathrm{T}};\mathrm{I}} \times \hat{\boldsymbol{h}}_m^{\mathrm{T;I}}\right) \cdot \boldsymbol{\nu}\,\mathrm{d}A \qquad (6.29)$$

that specifies the change of the radiation impedance matrix for all $m = \{1, \ldots, N\}$ and $n = \{1, \ldots, N\}$.

Table 6.5 Application of the reciprocity theorem.

Domain exterior to $\mathcal{S}_0 \cup \partial\mathcal{D}$		
Time-correlation	State (ΔT)	State (T)
Source	0	0
Field	$\{\Delta\hat{\boldsymbol{E}}^{\mathrm{T}}, \Delta\hat{\boldsymbol{H}}^{\mathrm{T}}\}$	$\{\hat{\boldsymbol{E}}^{\mathrm{T}}, \hat{\boldsymbol{H}}^{\mathrm{T}}\}$
Material	$\{\epsilon_0, \mu_0\}$	$\{\epsilon_0, \mu_0\}$

Norton's Network

The change in the radiation admittance matrix of the Norton network representation is in line with Section 5.1.2 described using the voltage-excited sensing-field EM wave field constituents. To this end, we take $\Delta\hat{V}_m^{\mathrm{T}} = 0$ for all $m = \{1, \dots, N\}$ and use Eq. (4.3) in the left-hand side of Eq. (6.27) and get

$$\sum_{m=1}^{N}\sum_{n=1}^{N} \hat{V}_m^{\mathrm{T}}(s)\Delta\hat{Y}_{m,n}^{\mathrm{T}}(s)\hat{V}_n^{\mathrm{T}}(s)$$
$$= \sum_{m=1}^{N}\sum_{n=1}^{N} \hat{V}_m^{\mathrm{T}}(s)\hat{V}_n^{\mathrm{T}}(s) \int_{\boldsymbol{x}\in\partial\mathcal{D}} \left(\hat{\boldsymbol{e}}_n^{\tilde{\mathrm{T}};\mathrm{V}} \times \hat{\boldsymbol{h}}_m^{\mathrm{T};\mathrm{V}} - \hat{\boldsymbol{e}}_m^{\mathrm{T};\mathrm{V}} \times \hat{\boldsymbol{h}}_n^{\tilde{\mathrm{T}};\mathrm{V}}\right) \cdot \boldsymbol{\nu}\,\mathrm{d}A \tag{6.30}$$

where we have used Eq. (5.18). Since the latter equality has to hold for arbitrary exciting voltages, we finally end up with

$$\Delta\hat{Y}_{m,n}^{\mathrm{T}}(s) = \int_{\boldsymbol{x}\in\partial\mathcal{D}} \left(\hat{\boldsymbol{e}}_n^{\tilde{\mathrm{T}};\mathrm{V}} \times \hat{\boldsymbol{h}}_m^{\mathrm{T};\mathrm{V}} - \hat{\boldsymbol{e}}_m^{\mathrm{T};\mathrm{V}} \times \hat{\boldsymbol{h}}_n^{\tilde{\mathrm{T}};\mathrm{V}}\right) \cdot \boldsymbol{\nu}\,\mathrm{d}A \tag{6.31}$$

that specifies the change of the radiation admittance matrix for all $m = \{1, \dots, N\}$ and $n = \{1, \dots, N\}$.

6.2.2 Analysis Based on the Reciprocity Theorem of the Time-Correlation Type

The present reciprocity analysis based on the reciprocity theorem of the time-correlation type partly follows the lines of the previous Section 6.2.1. Accordingly, we start with applying the reciprocity theorem to the unbounded domain exterior to the transmitting antenna system and to (ΔT) and (T) states (see Table 6.5).

Table 6.6 Application of the reciprocity theorem.

	Domain $\mathcal{A}$	
Time-correlation	State (ΔT)	State (T)
Source	0	0
Field	$\{\Delta\hat{\boldsymbol{E}}^{\mathrm{T}}, \Delta\hat{\boldsymbol{H}}^{\mathrm{T}}\}$	$\{\hat{\boldsymbol{E}}^{\mathrm{T}}, \hat{\boldsymbol{H}}^{\mathrm{T}}\}$
Material	$\{\hat{\underline{\eta}}, \hat{\underline{\zeta}}\}$	$\{\hat{\underline{\eta}}, \hat{\underline{\zeta}}\}$

Since both field states are causal, we may apply the expression similar to Eq. (1.45) and get

$$\int_{\boldsymbol{x}\in S_0\cup\partial\mathcal{D}} \left(\Delta\hat{\boldsymbol{E}}^{\mathrm{T}} \times \hat{\boldsymbol{H}}^{\mathrm{T}\circledast} + \hat{\boldsymbol{E}}^{\mathrm{T}\circledast} \times \Delta\hat{\boldsymbol{H}}^{\mathrm{T}}\right) \cdot \boldsymbol{\nu}\,\mathrm{d}A$$
$$= \left(\eta_0/8\pi^2\right) \int_{\boldsymbol{\xi}\in\Omega} \Delta\hat{\boldsymbol{E}}^{\mathrm{T};\infty}(\boldsymbol{\xi}, s) \cdot \hat{\boldsymbol{E}}^{\mathrm{T};\infty}(\boldsymbol{\xi}, -s)\mathrm{d}\Omega \tag{6.32}$$

In the following step, the reciprocity of the time-correlation type is applied to the bounded domain $\mathcal{A}$ and, again, to (ΔT) and (T) states (see Table 6.6). Using the generic form of the reciprocity theorem (1.36), we get the following relation:

$$\int_{\boldsymbol{x}\in S_0} \left(\Delta\hat{\boldsymbol{E}}^{\mathrm{T}} \times \hat{\boldsymbol{H}}^{\mathrm{T}\circledast} + \hat{\boldsymbol{E}}^{\mathrm{T}\circledast} \times \Delta\hat{\boldsymbol{H}}^{\mathrm{T}}\right) \cdot \boldsymbol{\nu}\,\mathrm{d}A$$
$$= \int_{\boldsymbol{x}\in S_1} \left(\Delta\hat{\boldsymbol{E}}^{\mathrm{T}} \times \hat{\boldsymbol{H}}^{\mathrm{T}\circledast} + \hat{\boldsymbol{E}}^{\mathrm{T}\circledast} \times \Delta\hat{\boldsymbol{H}}^{\mathrm{T}}\right) \cdot \boldsymbol{\nu}\,\mathrm{d}A$$
$$- \int_{\boldsymbol{x}\in\mathcal{A}} \left\{\Delta\hat{\boldsymbol{H}}^{\mathrm{T}} \cdot \left[\hat{\underline{\zeta}}^{\circledast} + \hat{\underline{\zeta}}^{\mathcal{T}}\right] \cdot \hat{\boldsymbol{H}}^{\mathrm{T}\circledast}\right.$$
$$\left. + \Delta\hat{\boldsymbol{E}}^{\mathrm{T}} \cdot \left[\hat{\underline{\eta}}^{\circledast} + \hat{\underline{\eta}}^{\mathcal{T}}\right] \cdot \hat{\boldsymbol{E}}^{\mathrm{T}\circledast}\right\}\mathrm{d}V \tag{6.33}$$

Here, the volume integral vanishes for the medium that is time-reverse self-adjoint in its EM properties (see Eqs. (1.37) and (1.38)). This includes loss-free antenna systems such as instantaneously reacting antennas described with symmetric tensors of electric permittivity and magnetic permeability (see Eqs. (1.48)–(1.49)), antenna systems made from PEC components (e.g., Figure 1.7), or, eventually, the combination of these cases. Furthermore, the surface integral over the terminal surface of the antenna system can be expressed in terms of Kirchhoff quantities, namely

Table 6.7 Application of the reciprocity theorem.

	Domain $\mathcal{D}$	
Time-correlation	State (T)	State (T)
Source	0	0
Field	$\{\hat{\boldsymbol{E}}^{\mathrm{T}}, \hat{\boldsymbol{H}}^{\mathrm{T}}\}$	$\{\hat{\boldsymbol{E}}^{\mathrm{T}}, \hat{\boldsymbol{H}}^{\mathrm{T}}\}$
Material	$\{\epsilon_0, \mu_0\}$	$\{\epsilon_0, \mu_0\}$

$$\int_{\boldsymbol{x}\in S_1} \left(\Delta\hat{\boldsymbol{E}}^{\mathrm{T}} \times \hat{\boldsymbol{H}}^{\mathrm{T}\circledast} + \hat{\boldsymbol{E}}^{\mathrm{T}\circledast} \times \Delta\hat{\boldsymbol{H}}^{\mathrm{T}}\right) \cdot \boldsymbol{\nu}\,\mathrm{d}A$$
$$\simeq \sum_{m=1}^{N} \left[\Delta\hat{V}_m^{\mathrm{T}}(s)\hat{I}_m^{\mathrm{T}}(-s) + \hat{V}_m^{\mathrm{T}}(-s)\Delta\hat{I}_m^{\mathrm{T}}(s)\right] \tag{6.34}$$

where we have used the fact that the excitation electric current is oriented along the outer unit normal vector $\boldsymbol{\nu}$ on the terminal surface S_1. Subsequently, the reciprocity theorem of the time-correlation type is applied to the domain occupied by the scatterer $\mathcal{D}$ and to one and the same state (T) (see Table 6.7 and Figure 6.2). Owing to the fact that state (T) is in $\mathcal{D}$ source-free and that the embedding is self-adjoint in its EM properties, we get

$$\int_{\boldsymbol{x}\in\partial\mathcal{D}} \left(\hat{\boldsymbol{E}}^{\mathrm{T}} \times \hat{\boldsymbol{H}}^{\mathrm{T}\circledast} + \hat{\boldsymbol{E}}^{\mathrm{T}\circledast} \times \hat{\boldsymbol{H}}^{\mathrm{T}}\right) \cdot \boldsymbol{\nu}\,\mathrm{d}A = 0 \tag{6.35}$$

In the final step, Eqs. (6.32)–(6.35) are combined together, which in view of Eq. (6.24) yields the desired reciprocity relation:

$$\sum_{m=1}^{N} \left[\Delta\hat{V}_m^{\mathrm{T}}(s)\hat{I}_m^{\mathrm{T}}(-s) + \hat{V}_m^{\mathrm{T}}(-s)\Delta\hat{I}_m^{\mathrm{T}}(s)\right]$$
$$= \left(\eta_0/8\pi^2\right) \int_{\boldsymbol{\xi}\in\Omega} \Delta\hat{\boldsymbol{E}}^{\mathrm{T};\infty}(\boldsymbol{\xi}, s) \cdot \hat{\boldsymbol{E}}^{\mathrm{T};\infty}(\boldsymbol{\xi}, -s)\mathrm{d}\Omega$$
$$- \int_{\boldsymbol{x}\in\partial\mathcal{D}} \left(\hat{\boldsymbol{E}}^{\mathrm{T}} \times \hat{\boldsymbol{H}}^{\mathrm{T}\circledast} + \hat{\boldsymbol{E}}^{\mathrm{T}\circledast} \times \hat{\boldsymbol{H}}^{\mathrm{T}}\right) \cdot \boldsymbol{\nu}\,\mathrm{d}A \tag{6.36}$$

In deriving the latter relation, we have assumed that the volume integral in Eq. (6.33) approaches zero, which applies to loss-free antenna systems. Equation (6.36) can be viewed as the time-correlation counterpart of the time-convolution reciprocity relation (6.27).

Thévenin's Network

Similarly to the previous section, the change in the radiation impedance matrix is expressed via the current-excited radiated-field EM wave constituents defined in Eqs. (4.23) and (5.16). Accordingly, we let $\Delta\hat{I}_m^{\rm T} = 0$ for all $m = \{1, \ldots, N\}$ in Eq. (6.36) and get

$$\Delta\hat{Z}^{\rm T}_{m,n}(s) = -\frac{1}{8\pi^2}\frac{s^2}{c_0^2}\frac{1}{\eta_0}\int_{\xi\in\Omega}\hat{\boldsymbol{e}}^{\rm T;I}_m(\xi,-s)\cdot\Delta\hat{\boldsymbol{e}}^{\rm T;I}_n(\xi,s){\rm d}\Omega - \int_{\boldsymbol{x}\in\partial D}\left(\hat{\boldsymbol{e}}^{\tilde{\rm T};{\rm I}}_n\times\hat{\boldsymbol{h}}^{\rm T;I\circledast}_m + \hat{\boldsymbol{e}}^{\rm T;I\circledast}_m\times\hat{\boldsymbol{h}}^{\tilde{\rm T};{\rm I}}_n\right)\cdot\boldsymbol{\nu}\,{\rm d}A \tag{6.37}$$

for all $m = \{1, \ldots, N\}$ and $n = \{1, \ldots, N\}$, where we have invoked the condition that the resulting identity obtained from Eq. (6.36) has to hold for arbitrary excitation currents. It is re-called here that the latter relation should be approached via the (limiting) real-FD, that is,

$$\Delta\hat{Z}^{\rm T}_{m,n}({\rm i}\omega) = \frac{1}{2\eta_0}\frac{1}{\lambda_0^2}\int_{\xi\in\Omega}\hat{\boldsymbol{e}}^{\rm T;I\circledast}_m(\xi,{\rm i}\omega)\cdot\Delta\hat{\boldsymbol{e}}^{\rm T;I}_n(\xi,{\rm i}\omega){\rm d}\Omega - \int_{\boldsymbol{x}\in\partial D}\left(\hat{\boldsymbol{e}}^{\tilde{\rm T};{\rm I}}_n\times\hat{\boldsymbol{h}}^{\rm T;I\circledast}_m + \hat{\boldsymbol{e}}^{\rm T;I\circledast}_m\times\hat{\boldsymbol{h}}^{\tilde{\rm T};{\rm I}}_n\right)\cdot\boldsymbol{\nu}\,{\rm d}A \tag{6.38}$$

for all $m = \{1, \ldots, N\}$ and $n = \{1, \ldots, N\}$, where $\lambda_0 = \omega/c_0 \in \mathbb{R}$ is the corresponding wavelength and symbol $\circledast$ has the meaning of complex conjugate.

Norton's Network

The change in the radiation admittance matrix is expressed here via the voltage-excited radiated-field EM wave constituents defined in Eqs. (5.10) and (5.18). Hence, we let $\Delta\hat{V}_m^{\rm T} = 0$ for all $m = \{1, \ldots, N\}$ in Eq. (6.36) and get

$$\Delta\hat{Y}^{\rm T}_{m,n}(s) = -\frac{\eta_0}{8\pi^2}\frac{s^2}{c_0^2}\int_{\xi\in\Omega}\hat{\boldsymbol{e}}^{\rm T;V}_m(\xi,-s)\cdot\Delta\hat{\boldsymbol{e}}^{\rm T;V}_n(\xi,s){\rm d}\Omega - \int_{\boldsymbol{x}\in\partial D}\left(\hat{\boldsymbol{e}}^{\tilde{\rm T};{\rm V}}_n\times\hat{\boldsymbol{h}}^{\rm T;V\circledast}_m + \hat{\boldsymbol{e}}^{\rm T;V\circledast}_m\times\hat{\boldsymbol{h}}^{\tilde{\rm T};{\rm V}}_n\right)\cdot\boldsymbol{\nu}\,{\rm d}A \tag{6.39}$$

for all $m = \{1, \ldots, N\}$ and $n = \{1, \ldots, N\}$, where we have invoked the condition that the resulting identity obtained from Eq. (6.36) has to hold for arbitrary excitation voltages. Again, upon approaching the latter via the real-FD, we get

$$\Delta\hat{Y}^{\rm T}_{m,n}({\rm i}\omega) = \frac{\eta_0}{2}\frac{1}{\lambda_0^2}\int_{\xi\in\Omega}\hat{\boldsymbol{e}}^{\rm T;V\circledast}_m(\xi,{\rm i}\omega)\cdot\Delta\hat{\boldsymbol{e}}^{\rm T;V}_n(\xi,{\rm i}\omega){\rm d}\Omega - \int_{\boldsymbol{x}\in\partial D}\left(\hat{\boldsymbol{e}}^{\tilde{\rm T};{\rm V}}_n\times\hat{\boldsymbol{h}}^{\rm T;V\circledast}_m + \hat{\boldsymbol{e}}^{\rm T;V\circledast}_m\times\hat{\boldsymbol{h}}^{\tilde{\rm T};{\rm V}}_n\right)\cdot\boldsymbol{\nu}\,{\rm d}A \tag{6.40}$$

for all $m = \{1, \ldots, N\}$ and $n = \{1, \ldots, N\}$.

Exercise

- Give an approximate expression for the contribution of a small spherical raindrop to the Thévenin voltage-source strength of a short-wire antenna (see Fig. 6.3). For simplicity, let us assume that the raindrop is located in the far-field region of the antenna.

Hint: Use Eqs. (6.14) and (6.8) with Eq. (5.16) for $N = 1$ to express the change of the Thévenin voltage-source strength in terms of the equivalent contrast electric current volume density:

$$\Delta\hat{V}^{\mathrm{G}}(s) = -\int_{\boldsymbol{x}\in D} \hat{\boldsymbol{e}}^{\mathrm{T;I}} \cdot \hat{\boldsymbol{J}}^{\tilde{\mathrm{R}}} \mathrm{d}V$$

where D is a spherical domain defined by the location of its center and by its (relatively small) radius $\varrho > 0$. Now, under the assumption that the sensing field $\hat{\boldsymbol{e}}^{\mathrm{T;I}}$ is uniform over D, we write

$$\Delta\hat{V}^{\mathrm{G}}(s) \simeq -\hat{\boldsymbol{e}}^{\mathrm{T;I}} \cdot \int_{\boldsymbol{x}\in D} \hat{\boldsymbol{J}}^{\tilde{\mathrm{R}}} \mathrm{d}V$$

where the volume integral of the contrast electric current density is associated with $s\hat{\boldsymbol{p}}$ through the electric dipole moment $\hat{\boldsymbol{p}}$. The latter is, in virtue of the Born-like approximation, expressed as

$$\hat{\boldsymbol{p}} \overset{\mathrm{B}}{\simeq} \epsilon_0 \hat{I}^{\mathrm{T}}(s)\, \underline{\hat{\boldsymbol{\alpha}}}^{\mathrm{E}} \cdot \hat{\boldsymbol{e}}^{\mathrm{T;I}}$$

using the polarizability tensor of the electric type, $\underline{\hat{\boldsymbol{\alpha}}}^{\mathrm{E}}$, for which we use the static-field approximation applying to a homogeneous dielectric sphere ([46], Eq. (3.126)):

$$\underline{\hat{\boldsymbol{\alpha}}}^{\mathrm{E}} = 4\pi\varrho^3[(\epsilon_{\mathrm{r}} - 1)/(\epsilon_{\mathrm{r}} + 2)]\underline{\boldsymbol{I}}$$

where ϵ_{r} is its relative permittivity. Accordingly, we then write

$$\begin{aligned}\Delta\hat{V}^{\mathrm{G}}(s) &\simeq -s\epsilon_0 \hat{I}^{\mathrm{T}}(s)\, \hat{\boldsymbol{e}}^{\mathrm{T;I}} \cdot \underline{\hat{\boldsymbol{\alpha}}}^{\mathrm{E}} \cdot \hat{\boldsymbol{e}}^{\mathrm{T;I}} \\ &= -4\pi s\epsilon_0 \hat{I}^{\mathrm{T}}(s)\, \varrho^3[(\epsilon_{\mathrm{r}} - 1)/(\epsilon_{\mathrm{r}} + 2)]\hat{\boldsymbol{e}}^{\mathrm{T;I}} \cdot \hat{\boldsymbol{e}}^{\mathrm{T;I}}\end{aligned} \tag{6.41}$$

The radiated-field EM field constituents admit the far-field expansion that can be expressed as (see Eqs. (4.10) with (4.23) and (5.16))

$$\hat{\boldsymbol{e}}^{\mathrm{T;I}} = s\mu_0 \hat{\boldsymbol{\mathcal{e}}}^{\mathrm{T;I}} \exp(-s|\boldsymbol{x}|/c_0)(4\pi|\boldsymbol{x}|)^{-1}\left[1 + O(|\boldsymbol{x}|^{-1})\right] \tag{6.42}$$

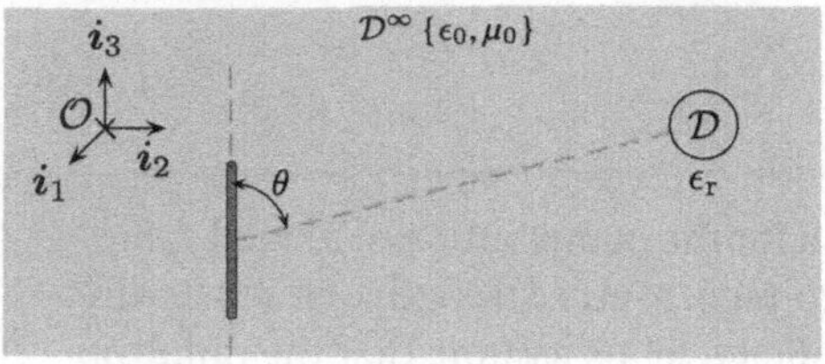

Figure 6.3 Short-wire antenna interaction with a spherical raindrop.

as $|\boldsymbol{x}| \to \infty$ and the effective length for a short wire carrying the uniform current distribution reads

$$\hat{\boldsymbol{\ell}}^{\mathrm{T;I}}(\boldsymbol{\xi}, s) = (\boldsymbol{\xi}\boldsymbol{\xi} - \underline{\boldsymbol{I}}) \cdot \boldsymbol{\ell}$$

in which $\boldsymbol{\ell}$ is the vectorial length of the wire. Finally, after collecting the results, we end up with

$$\Delta\hat{V}^{\mathrm{G}}(s) \simeq -4\pi\varrho^3 s\mu_0 \hat{I}^{\mathrm{T}}(s)[(\epsilon_{\mathrm{r}} - 1)/(\epsilon_{\mathrm{r}} + 2)](s^2\ell^2/c_0^2)\sin^2(\theta) \times [\exp(-s|\boldsymbol{x}|/c_0)/(4\pi|\boldsymbol{x}|)]^2 \quad (6.43)$$

where $\ell = |\boldsymbol{\ell}|$ is the length of the wire. Upon neglecting multiple scattering effects, we may apply the latter expression to integrate the contributions from a cluster of raindrops and, consequently, use the result to evaluate the corresponding power received by the (matched) antenna. For more details on the subject, we refer the reader to Sec. 6.8 of Ref. [9].

7

EM Coupling Between Two Multiport Antenna Systems

Perhaps the most famous consequence of reciprocity in EM theory is that the transmitting and receiving patterns of an antenna system are, subject to certain conditions, fully equivalent. This property is widely applied to a typical antenna pattern measurement configuration in which the (test) antenna, whose EM radiation characteristics are measured, acts as a receiver of the signal transmitted from the (stationary) referential antenna ([45], Ch. 13). A necessary condition for such measurements to be successful is the property of reciprocity of the corresponding transfer impedance ([38] Sec. 2.13). From this reason, it is of high practical importance to carefully analyze the remote interaction of two antenna systems. This is exactly the main purpose of this chapter, where the transfer impedance matrix and its properties are analyzed with the aid of reciprocity theorems of both the time-convolution and time-correlation types.

7.1 Description of the Problem Configuration

We shall next distinguish between two scenarios shown in Figure 7.1, each consisting of one transmitting and one receiving antenna located in the linear, homogeneous, and isotropic embedding $\mathcal{D}^{\infty}$. In the analysis that follows, antenna (A) is represented as an M-port system and antenna (B) as a N-port system. The antenna systems (A) and (B) consist of a medium whose EM properties are specified by the (tensorial) transverse admittance and the longitudinal impedance denoted by $\{\underline{\hat{\eta}}^{A}, \underline{\hat{\zeta}}^{A}\}$ and $\{\underline{\hat{\eta}}^{B}, \underline{\hat{\zeta}}^{B}\}$ (see Eqs. (1.23) and (1.24)). The analyzed antenna configurations may also contain (perfectly) conducting parts (see Section 1.5 for details). In state (BA), antenna (A) operates as a radiator of EM wave fields and antenna (B) acts as a loaded scatterer, while in state (AB) their roles are mutually interchanged.

In accordance with standard network theory ([7], Ch. 8), EM coupling between two such linear and time-invariant antenna multiport systems can be described in terms of network parameters. Without loss of generality, we shall next choose the so-called (open-circuit) impedance parameters. Accordingly, in state (BA), the voltages $\hat{V}_m^{A;\mathrm{T}}$ and $\hat{V}_n^{B;\mathrm{R}}$ at accessible ports of the transmit-

Electromagnetic Reciprocity in Antenna Theory, First Edition. Martin Stumpf.

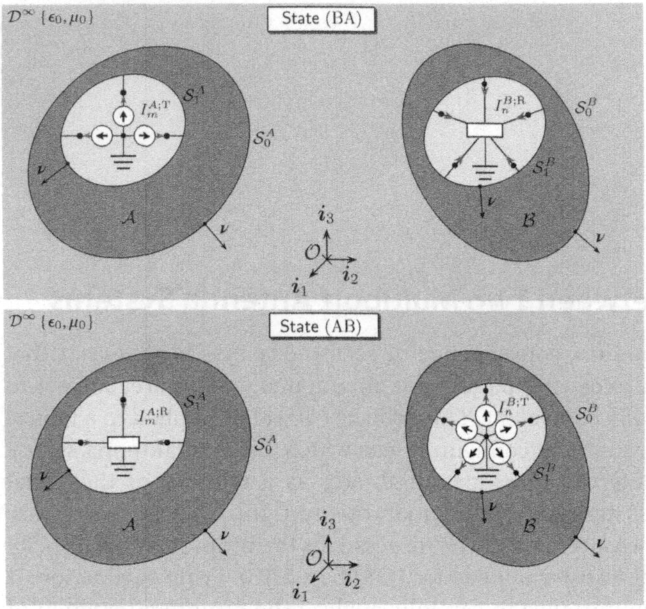

Figure 7.1 State (BA) with transmitting antenna (A) and receiving antenna (B) and state (AB) with transmitting antenna (B) and receiving antenna (A). Antenna (A) has $M = 3$ ports and antenna (B) has $N = 5$ ports.

ting antenna (A) and the receiving antenna (B), respectively, are expressed as the linear combination of the corresponding electric currents $\hat{I}_m^{A;\mathrm{T}}$ and $\hat{I}_n^{B;\mathrm{R}}$, namely,

$$\hat{\boldsymbol{V}}^{A;\mathrm{T}}(s) = \underline{\hat{\boldsymbol{Z}}}^{A;\mathrm{T}}(s) \cdot \hat{\boldsymbol{I}}^{A;\mathrm{T}}(s) + \underline{\hat{\boldsymbol{Z}}}^{AB}(s) \cdot \hat{\boldsymbol{I}}^{B;\mathrm{R}}(s) \tag{7.1}$$

$$\hat{\boldsymbol{V}}^{B;\mathrm{R}}(s) = \underline{\hat{\boldsymbol{Z}}}^{BA}(s) \cdot \hat{\boldsymbol{I}}^{A;\mathrm{T}}(s) + \underline{\hat{\boldsymbol{Z}}}^{B;\mathrm{L}}(s) \cdot \hat{\boldsymbol{I}}^{B;\mathrm{R}}(s) \tag{7.2}$$

where the impedance matrices have the following meaning:

- $\underline{\hat{\boldsymbol{Z}}}^{A;\mathrm{T}}$ is the $[M \times M]$ (radiation) impedance matrix describing the local interactions of the M-port transmitting antenna system (A).
- $\underline{\hat{\boldsymbol{Z}}}^{B;\mathrm{L}}$ is the $[N \times N]$ (load) impedance matrix describing the local interactions of the N-port receiving antenna system (B).
- $\underline{\hat{\boldsymbol{Z}}}^{AB}$ is the $[M \times N]$ (transfer) impedance matrix describing the remote interactions from N-port receiving antenna system (B) to M-port transmitting antenna system (A).
- $\underline{\hat{\boldsymbol{Z}}}^{BA}$ is the $[N \times M]$ (transfer) impedance matrix describing the remote interactions from M-port transmitting antenna system (A) to N-port receiving antenna system (B).

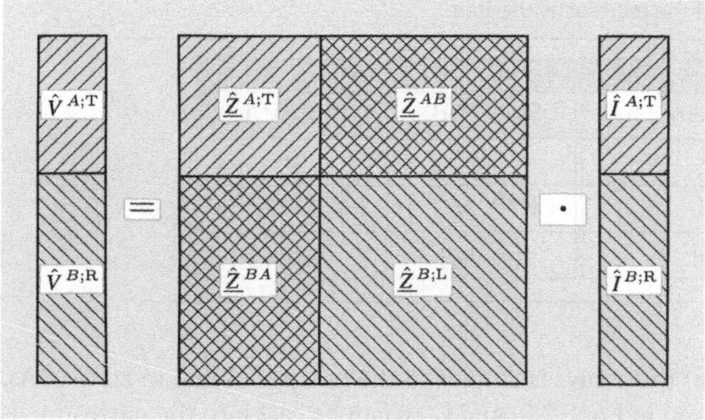

Figure 7.2 Partitioned impedance matrix description of state (BA).

In line with Sec. 30.4 of Ref. [16], such interaction can be described using one $[(M+N)\times(M+N)]$ partitioned impedance matrix that is broken into the impedance submatrices whose meaning is given above (see Figure 7.2). The block matrix system can be then written as

$$\hat{V}_p^{A;\mathrm{T}}(s) = \sum_{q=1}^{M} \hat{Z}_{p,q}^{A;\mathrm{T}}(s)\hat{I}_q^{A;\mathrm{T}}(s) + \sum_{q=M+1}^{M+N} \hat{Z}_{p,q}^{AB}(s)\hat{I}_q^{B;\mathrm{R}}(s) \tag{7.3}$$

for $p = \{1, \dots, M\}$

$$\hat{V}_p^{B;\mathrm{R}}(s) = \sum_{n=1}^{M} \hat{Z}_{p,q}^{BA}(s)\hat{I}_q^{A;\mathrm{T}}(s) + \sum_{q=M+1}^{M+N} \hat{Z}_{p,q}^{B;\mathrm{L}}(s)\hat{I}_q^{B;\mathrm{R}}(s) \tag{7.4}$$

for $p = \{M+1, \dots, M+N\}$

where subscripts $\{p, q\}$ refer to the partitioned arrays. Accordingly, the elements of the transfer matrix $\hat{\underline{Z}}^{AB}$, for instance, can be specified either as $\hat{Z}_{p,q}^{AB}$ for $p = \{1, \dots, M\}$ and $q = \{M+1, \dots, M+N\}$ in the partitioned matrix index notation, or, equivalently, as $\hat{Z}_{m,n}^{AB}$ for $m = \{1, \dots, M\}$ and $n = \{1, \dots, N\}$ using its local indices $\{m, n\}$. In the description that follows, we shall prefer the local index notation, while keeping the block matrix format in mind, in particular, for computer implementation purposes.

Similar impedance parameters can also be defined for state (AB), where antenna (B) operates as a transmitter, while antenna (A) acts as a receiver. Upon interchanging (A) and (B) in Eqs. (7.1) and (7.2), we obtain

$$\hat{\boldsymbol{V}}^{B;\mathrm{T}}(s) = \hat{\underline{\boldsymbol{Z}}}^{B;\mathrm{T}}(s) \cdot \hat{\boldsymbol{I}}^{B;\mathrm{T}}(s) + \hat{\underline{\boldsymbol{Z}}}^{BA}(s) \cdot \hat{\boldsymbol{I}}^{A;\mathrm{R}}(s) \tag{7.5}$$

$$\hat{\boldsymbol{V}}^{A;\mathrm{R}}(s) = \hat{\underline{\boldsymbol{Z}}}^{AB}(s) \cdot \hat{\boldsymbol{I}}^{B;\mathrm{T}}(s) + \hat{\underline{\boldsymbol{Z}}}^{A;\mathrm{L}}(s) \cdot \hat{\boldsymbol{I}}^{A;\mathrm{R}}(s) \tag{7.6}$$

Table 7.1 Application of the reciprocity theorem.

Domain exterior to $\mathcal{S}_0^A \cup \mathcal{S}_0^B$		
Time-convolution	State (AB)	State (BA)
Source	0	0
Field	$\{\hat{\boldsymbol{E}}^{\text{AB}}, \hat{\boldsymbol{H}}^{\text{AB}}\}$	$\{\hat{\boldsymbol{E}}^{\text{BA}}, \hat{\boldsymbol{H}}^{\text{BA}}\}$
Material	$\{\epsilon_0, \mu_0\}$	$\{\epsilon_0, \mu_0\}$

where the meaning of the arrays is similar to the one pertaining to state (BA). Again, the system given in Eqs. (7.5) and (7.6) can be cast into the partitioned-matrix form that is similar to the one described using Eqs. (7.3) and (7.4) (cf. Figure 7.2).

7.2 Analysis Based on the Reciprocity Theorem of the Time-Convolution Type

In this section, the mutual coupling between the two multiport antenna systems is analyzed with the aid of the reciprocity theorem of the time-convolution type (see Section 1.4.1). In the first step, the reciprocity theorem is applied to the unbounded domain exterior to both interacting antenna systems and to the total EM wave fields in the analyzed situations (see Table 7.1).

Using the generic form of the reciprocity theorem (1.30) in combination with the conclusions drawn in Section 1.4.3, we end up with

$$\int_{\boldsymbol{x}\in\mathcal{S}_0^A} \left(\hat{\boldsymbol{E}}^{\text{AB}} \times \hat{\boldsymbol{H}}^{\text{BA}} - \hat{\boldsymbol{E}}^{\text{BA}} \times \hat{\boldsymbol{H}}^{\text{AB}}\right) \cdot \boldsymbol{\nu}\,\mathrm{d}A = \int_{\boldsymbol{x}\in\mathcal{S}_0^B} \left(\hat{\boldsymbol{E}}^{\text{BA}} \times \hat{\boldsymbol{H}}^{\text{AB}} - \hat{\boldsymbol{E}}^{\text{AB}} \times \hat{\boldsymbol{H}}^{\text{BA}}\right) \cdot \boldsymbol{\nu}\,\mathrm{d}A \tag{7.7}$$

where we have taken into account the self-adjoint properties of the embedding. If, in addition, antenna $(\mathcal{A})$ is reciprocal in its EM behavior, that is,

$$\underline{\hat{\eta}}^A(\boldsymbol{x}, s) = (\underline{\hat{\eta}}^A)^{\mathcal{T}}(\boldsymbol{x}, s) \tag{7.8}$$

$$\underline{\hat{\zeta}}^A(\boldsymbol{x}, s) = (\underline{\hat{\zeta}}^A)^{\mathcal{T}}(\boldsymbol{x}, s) \tag{7.9}$$

for all $\boldsymbol{x} \in \mathcal{A}$, the surface integral over its bounding surface $\mathcal{S}_0^A$ is equivalent to the one taken on its terminal surface $\mathcal{S}_1^A$ that can be expressed using the corresponding Kirchhoff circuit quantities (see Section 1.5.2), specifically

$$\int_{\boldsymbol{x}\in\mathcal{S}_0^A} \left(\hat{\boldsymbol{E}}^{\mathrm{AB}} \times \hat{\boldsymbol{H}}^{\mathrm{BA}} - \hat{\boldsymbol{E}}^{\mathrm{BA}} \times \hat{\boldsymbol{H}}^{\mathrm{AB}}\right) \cdot \boldsymbol{\nu}\,\mathrm{d}A$$
$$= \int_{\boldsymbol{x}\in\mathcal{S}_1^A} \left(\hat{\boldsymbol{E}}^{\mathrm{AB}} \times \hat{\boldsymbol{H}}^{\mathrm{BA}} - \hat{\boldsymbol{E}}^{\mathrm{BA}} \times \hat{\boldsymbol{H}}^{\mathrm{AB}}\right) \cdot \boldsymbol{\nu}\,\mathrm{d}A$$
$$\simeq \sum_{m=1}^{M} [\hat{V}_m^{A;\mathrm{R}}(s)\hat{I}_m^{A;\mathrm{T}}(s) + \hat{V}_m^{A;\mathrm{T}}(s)\hat{I}_m^{A;\mathrm{R}}(s)] \tag{7.10}$$

where we have accounted for the orientation of the electric currents with respect to $\boldsymbol{\nu}$ along $\mathcal{S}_1^A$. Hence, combination of Eqs. (7.7)–(7.10) results in

$$\sum_{m=1}^{M} [\hat{V}_m^{A;\mathrm{R}}(s)\hat{I}_m^{A;\mathrm{T}}(s) + \hat{V}_m^{A;\mathrm{T}}(s)\hat{I}_m^{A;\mathrm{R}}(s)]$$
$$= \int_{\boldsymbol{x}\in\mathcal{S}_0^B} \left(\hat{\boldsymbol{E}}^{\mathrm{BA}} \times \hat{\boldsymbol{H}}^{\mathrm{AB}} - \hat{\boldsymbol{E}}^{\mathrm{AB}} \times \hat{\boldsymbol{H}}^{\mathrm{BA}}\right) \cdot \boldsymbol{\nu}\,\mathrm{d}A \tag{7.11}$$

A special case arises when the (receiving) antenna (A) in state (AB) is left open-circuited at all of its ports. Consequently, Eq. (7.6) leads to

$$\hat{V}_m^{A;\mathrm{R}}(s) = \sum_{n=1}^{N} \hat{Z}_{m,n}^{AB}(s)\hat{I}_n^{B;\mathrm{T}}(s)|_{\hat{I}_m^{A;\mathrm{R}}=0} \tag{7.12}$$

for all $m = \{1, \dots, M\}$ and we end up with

$$\hat{Z}_{m,n}^{AB}(s) = \int_{\boldsymbol{x}\in\mathcal{S}_0^B} \left(\hat{\boldsymbol{e}}_m^{\mathrm{BA;I}} \times \hat{\boldsymbol{h}}_n^{\mathrm{AB;I}} - \hat{\boldsymbol{e}}_n^{\mathrm{AB;I}} \times \hat{\boldsymbol{h}}_m^{\mathrm{BA;I}}\right) \cdot \boldsymbol{\nu}\,\mathrm{d}A \tag{7.13}$$

for all $m = \{1, \dots, M\}$ and $n = \{1, \dots, N\}$, where we have introduced the electric-current-excited radiated EM field wave constituents according to Eq. (5.16) and invoked the condition that the resulting identity has to hold for arbitrary values of $\hat{I}_n^{B;\mathrm{T}}$ and $\hat{I}_m^{A;\mathrm{T}}$ for all $m = \{1, \dots, M\}$ and $n = \{1, \dots, N\}$. Finally, recall that the resulting Eq. (7.13) describing the transfer impedance matrix of the interaction from antenna system (B) to antenna system (A) has been derived under the assumption that antenna (A) and the embedding are reciprocal in their EM constitutive properties, whatever the reciprocity behavior of antenna (B).

Similarly, in the case that antenna (B) is reciprocal in its EM behavior, that is,

$$\underline{\hat{\eta}}^{B}(\boldsymbol{x},s) = (\underline{\hat{\eta}}^{B})^{\mathcal{T}}(\boldsymbol{x},s) \tag{7.14}$$

$$\underline{\hat{\zeta}}^{B}(\boldsymbol{x},s) = (\underline{\hat{\zeta}}^{B})^{\mathcal{T}}(\boldsymbol{x},s) \tag{7.15}$$

the starting reciprocity relation (7.7) results in

$$\sum_{n=1}^{N}\left[\hat{V}_n^{B;\mathrm{R}}(s)\hat{I}_n^{B;\mathrm{T}}(s) + \hat{V}_n^{B;\mathrm{T}}(s)\hat{I}_n^{B;\mathrm{R}}(s)\right] = \int_{\boldsymbol{x}\in\mathcal{S}_0^A}\left(\hat{\boldsymbol{E}}^{\mathrm{AB}}\times\hat{\boldsymbol{H}}^{\mathrm{BA}} - \hat{\boldsymbol{E}}^{\mathrm{BA}}\times\hat{\boldsymbol{H}}^{\mathrm{AB}}\right)\cdot\boldsymbol{\nu}\,\mathrm{d}A \tag{7.16}$$

Again, if the receiving antenna (B) in state (BA) is open-circuited at all of its terminal ports, Eq. (7.2) gives

$$\hat{V}_n^{B;\mathrm{R}}(s) = \sum_{m=1}^{M}\hat{Z}_{n,m}^{BA}(s)\hat{I}_m^{A;\mathrm{T}}(s)|_{\hat{I}_n^{B;\mathrm{R}}=0} \tag{7.17}$$

for all $n = \{1, \dots, N\}$ and we end up with

$$\hat{Z}_{n,m}^{BA}(s) = \int_{\boldsymbol{x}\in\mathcal{S}_0^A}\left(\hat{\boldsymbol{e}}_n^{\mathrm{AB;I}}\times\hat{\boldsymbol{h}}_m^{\mathrm{BA;I}} - \hat{\boldsymbol{e}}_m^{\mathrm{BA;I}}\times\hat{\boldsymbol{h}}_n^{\mathrm{AB;I}}\right)\cdot\boldsymbol{\nu}\,\mathrm{d}A \tag{7.18}$$

for all $m = \{1, \dots, M\}$ and $n = \{1, \dots, N\}$, where we have introduced the electric-current-excited radiated EM field wave constituents according to Eq. (5.16) and invoked the condition that the resulting identity has to hold for arbitrary values of $\hat{I}_n^{B;\mathrm{T}}$ and $\hat{I}_m^{A;\mathrm{T}}$ for all $m = \{1, \dots, M\}$ and $n = \{1, \dots, N\}$. The resulting Eq. (7.18) describing the transfer impedance matrix of the interaction from antenna system (A) to antenna system (B) has been derived under the assumption that antenna (B) and the embedding are reciprocal in their EM constitutive properties, whatever the reciprocity behavior of antenna (A).

In the case that conditions (7.8) and (7.9) as well as (7.14) and (7.15) do apply simultaneously, the starting reciprocity relation (7.7) yields the following identity:

$$\sum_{m=1}^{M}\left[\hat{V}_m^{A;\mathrm{R}}(s)\hat{I}_m^{A;\mathrm{T}}(s) + \hat{V}_m^{A;\mathrm{T}}(s)\hat{I}_m^{A;\mathrm{R}}(s)\right] = \sum_{n=1}^{N}\left[\hat{V}_n^{B;\mathrm{R}}(s)\hat{I}_n^{B;\mathrm{T}}(s) + \hat{V}_n^{B;\mathrm{T}}(s)\hat{I}_n^{B;\mathrm{R}}(s)\right] \tag{7.19}$$

Table 7.2 Application of the reciprocity theorem.

Domain exterior to $\mathcal{S}_0^A \cup \mathcal{S}_0^B$		
Time-correlation	State (AB)	State (BA)
Source	0	0
Field	$\{\hat{\boldsymbol{E}}^{\mathrm{AB}}, \hat{\boldsymbol{H}}^{\mathrm{AB}}\}$	$\{\hat{\boldsymbol{E}}^{\mathrm{BA}}, \hat{\boldsymbol{H}}^{\mathrm{BA}}\}$
Material	$\{\epsilon_0, \mu_0\}$	$\{\epsilon_0, \mu_0\}$

which for open-circuited antenna systems, that is, $\hat{I}_m^{A;\mathrm{R}} = 0$ for all $m = \{1, \dots, M\}$ and $\hat{I}_n^{B;\mathrm{R}} = 0$ for all $n = \{1, \dots, N\}$, boils down to

$$\hat{Z}_{m,n}^{AB}(s) = \hat{Z}_{n,m}^{BA}(s) \tag{7.20}$$

for all $m = \{1, \dots, M\}$ and $n = \{1, \dots, N\}$, upon invoking the condition that the resulting equality has to hold for arbitrary exciting electric currents $\hat{I}_m^{A;\mathrm{T}}$ and $\hat{I}_n^{B;\mathrm{T}}$.

7.3 Analysis Based on the Reciprocity Theorem of the Time-Correlation Type

In this section, the remote interaction between the two multiport antenna systems is analyzed with the aid of the reciprocity theorem of the time-correlation type (see Section 1.4.2). To this end, the reciprocity theorem is first applied to the unbounded domain exterior to both interacting antenna systems and to the total EM wave fields in the analyzed situations (see Table 7.2).

Making use of the generic form of the reciprocity theorem (1.36) in combination with the conclusions drawn in Section 1.4.3, we end up with

$$\begin{aligned}
&\int_{\boldsymbol{x}\in\mathcal{S}_0^A} \left(\hat{\boldsymbol{E}}^{\mathrm{AB}} \times \hat{\boldsymbol{H}}^{\mathrm{BA}\circledast} + \hat{\boldsymbol{E}}^{\mathrm{BA}\circledast} \times \hat{\boldsymbol{H}}^{\mathrm{AB}}\right) \cdot \boldsymbol{\nu}\,\mathrm{d}A \\
&= (\eta_0/8\pi^2) \int_{\boldsymbol{\xi}\in\Omega} \hat{\boldsymbol{E}}^{\mathrm{AB};\infty}(\boldsymbol{\xi}, s) \cdot \hat{\boldsymbol{E}}^{\mathrm{BA};\infty}(\boldsymbol{\xi}, -s)\mathrm{d}\Omega \\
&\quad - \int_{\boldsymbol{x}\in\mathcal{S}_0^B} \left(\hat{\boldsymbol{E}}^{\mathrm{BA}\circledast} \times \hat{\boldsymbol{H}}^{\mathrm{AB}} + \hat{\boldsymbol{E}}^{\mathrm{AB}} \times \hat{\boldsymbol{H}}^{\mathrm{BA}\circledast}\right) \cdot \boldsymbol{\nu}\,\mathrm{d}A
\end{aligned} \tag{7.21}$$

by virtue of the self-adjointness of the embedding. For a class of antenna systems (A) whose constitutive properties are time-reverse self-adjoint, that is,

$$\underline{\hat{\eta}}^A(\boldsymbol{x},-s) = -(\underline{\hat{\eta}}^A)^T(\boldsymbol{x},s) \tag{7.22}$$

$$\underline{\hat{\zeta}}^A(\boldsymbol{x},-s) = -(\underline{\hat{\zeta}}^A)^T(\boldsymbol{x},s) \tag{7.23}$$

which applies to loss-free antenna systems (see Section 1.5.1), the surface integral over the bounding surface $\mathcal{S}_0^A$ can be directly expressed in terms of the corresponding Kirchhoff circuit quantities (see Section 1.5.2), specifically

$$\begin{aligned}
&\int_{\boldsymbol{x}\in\mathcal{S}_0^A}\left(\hat{\boldsymbol{E}}^{\mathrm{AB}}\times\hat{\boldsymbol{H}}^{\mathrm{BA}\circledast}+\hat{\boldsymbol{E}}^{\mathrm{BA}\circledast}\times\hat{\boldsymbol{H}}^{\mathrm{AB}}\right)\cdot\boldsymbol{\nu}\,\mathrm{d}A\\
&=\int_{\boldsymbol{x}\in\mathcal{S}_1^A}\left(\hat{\boldsymbol{E}}^{\mathrm{AB}}\times\hat{\boldsymbol{H}}^{\mathrm{BA}\circledast}+\hat{\boldsymbol{E}}^{\mathrm{BA}\circledast}\times\hat{\boldsymbol{H}}^{\mathrm{AB}}\right)\cdot\boldsymbol{\nu}\,\mathrm{d}A\\
&\simeq\sum_{m=1}^{M}\left[\hat{V}_m^{A;\mathrm{R}}(s)\hat{I}_m^{A;\mathrm{T}}(-s)-\hat{V}_m^{A;\mathrm{T}}(-s)\hat{I}_m^{A;\mathrm{R}}(s)\right]
\end{aligned} \tag{7.24}$$

where we have adhered to the orientation of the electric currents on the terminal surface $\mathcal{S}_1^A$ (see Figure 7.2). Consequently, upon combining Eqs. (7.21)–(7.24), we arrive at

$$\begin{aligned}
&\sum_{m=1}^{M}\left[\hat{V}_m^{A;\mathrm{R}}(s)\hat{I}_m^{A;\mathrm{T}}(-s)-\hat{V}_m^{A;\mathrm{T}}(-s)\hat{I}_m^{A;\mathrm{R}}(s)\right]\\
&=(\eta_0/8\pi^2)\int_{\boldsymbol{\xi}\in\Omega}\hat{\boldsymbol{E}}^{\mathrm{AB};\infty}(\boldsymbol{\xi},s)\cdot\hat{\boldsymbol{E}}^{\mathrm{BA};\infty}(\boldsymbol{\xi},-s)\mathrm{d}\Omega\\
&\quad-\int_{\boldsymbol{x}\in\mathcal{S}_0^B}\left(\hat{\boldsymbol{E}}^{\mathrm{BA}\circledast}\times\hat{\boldsymbol{H}}^{\mathrm{AB}}+\hat{\boldsymbol{E}}^{\mathrm{AB}}\times\hat{\boldsymbol{H}}^{\mathrm{BA}\circledast}\right)\cdot\boldsymbol{\nu}\,\mathrm{d}A
\end{aligned} \tag{7.25}$$

If the receiving antenna (A) is in state (AB) open-circuited, we may use Eq. (7.12) and get

$$\begin{aligned}
\hat{Z}_{m,n}^{AB}(s)&=-\frac{1}{8\pi^2}\frac{s^2}{c_0^2}\frac{1}{\eta_0}\int_{\boldsymbol{\xi}\in\Omega}\hat{\boldsymbol{e}}_m^{\mathrm{BA;I}}(\boldsymbol{\xi},-s)\cdot\hat{\boldsymbol{e}}_n^{\mathrm{AB;I}}(\boldsymbol{\xi},s)\mathrm{d}\Omega\\
&\quad-\int_{\boldsymbol{x}\in\mathcal{S}_0^B}\left(\hat{\boldsymbol{e}}_m^{\mathrm{BA;I}\circledast}\times\hat{\boldsymbol{h}}_n^{\mathrm{AB;I}}+\hat{\boldsymbol{e}}_n^{\mathrm{AB;I}}\times\hat{\boldsymbol{h}}_m^{\mathrm{BA;I}\circledast}\right)\cdot\boldsymbol{\nu}\,\mathrm{d}A
\end{aligned} \tag{7.26}$$

for all $m = \{1, \dots, N\}$ and $n = \{1, \dots, N\}$, where we have introduced the electric-current-excited radiated EM field wave constituents according to Eqs. (5.16) and (4.23) and, subsequently, invoked the condition that the resulting identity has to hold for arbitrary exciting electric currents $\hat{I}_m^{A;\mathrm{T}}$ and $\hat{I}_n^{B;\mathrm{T}}$. Such a correlation-type reciprocity relation should be approached via the (limiting) real-FD, which leads to

$$\hat{Z}^{AB}_{m,n}(\mathrm{i}\omega) = \frac{1}{2\eta_0}\frac{1}{\lambda_0^2}\int_{\xi\in\Omega} \hat{\boldsymbol{e}}^{\mathrm{BA;I}\circledast}_m(\boldsymbol{\xi},\mathrm{i}\omega)\cdot\hat{\boldsymbol{e}}^{\mathrm{AB;I}}_n(\boldsymbol{\xi},\mathrm{i}\omega)\mathrm{d}\Omega - \int_{\boldsymbol{x}\in S_0^B}\left(\hat{\boldsymbol{e}}^{\mathrm{BA;I}\circledast}_m\times\hat{\boldsymbol{h}}^{\mathrm{AB;I}}_n + \hat{\boldsymbol{e}}^{\mathrm{AB;I}}_n\times\hat{\boldsymbol{h}}^{\mathrm{BA;I}\circledast}_m\right)\cdot\boldsymbol{\nu}\,\mathrm{d}A \tag{7.27}$$

for all $m = \{1, \dots, N\}$ and $n = \{1, \dots, N\}$, where $\lambda_0 = \omega/c_0 \in \mathbb{R}$ is the corresponding wavelength and symbol $\circledast$ has the meaning of complex conjugate.

Similarly, if the medium of antenna (B) is time-reverse self-adjoint, that is,

$$\hat{\underline{\eta}}^B(\boldsymbol{x},-s) = -(\hat{\underline{\eta}}^B)^{\mathcal{T}}(\boldsymbol{x},s) \tag{7.28}$$

$$\hat{\underline{\zeta}}^B(\boldsymbol{x},-s) = -(\hat{\underline{\zeta}}^B)^{\mathcal{T}}(\boldsymbol{x},s) \tag{7.29}$$

the starting reciprocity relation (7.21) yields

$$\sum_{n=1}^{N}\left[\hat{V}^{B;\mathrm{R}}_n(-s)\hat{I}^{B;\mathrm{T}}_n(s) - \hat{V}^{B;\mathrm{T}}_n(s)\hat{I}^{B;\mathrm{R}}_n(-s)\right] = \left(\eta_0/8\pi^2\right)\int_{\xi\in\Omega}\hat{\boldsymbol{E}}^{\mathrm{AB};\infty}(\boldsymbol{\xi},s)\cdot\hat{\boldsymbol{E}}^{\mathrm{BA};\infty}(\boldsymbol{\xi},-s)\mathrm{d}\Omega - \int_{\boldsymbol{x}\in S_0^A}\left(\hat{\boldsymbol{E}}^{\mathrm{AB}}\times\hat{\boldsymbol{H}}^{\mathrm{BA}\circledast} + \hat{\boldsymbol{E}}^{\mathrm{BA}\circledast}\times\hat{\boldsymbol{H}}^{\mathrm{AB}}\right)\cdot\boldsymbol{\nu}\,\mathrm{d}A \tag{7.30}$$

for loss-free antenna systems (B). If the receiving antenna (B) in state (BA) is open-circuited at its accessible ports, the use of Eq. (7.17) in the reciprocity relation (7.30) results in

$$\hat{Z}^{BA}_{n,m}(-s) = -\frac{1}{8\pi^2}\frac{s^2}{c_0^2}\frac{1}{\eta_0}\int_{\xi\in\Omega}\hat{\boldsymbol{e}}^{\mathrm{BA;I}}_m(\boldsymbol{\xi},-s)\cdot\hat{\boldsymbol{e}}^{\mathrm{AB;I}}_n(\boldsymbol{\xi},s)\mathrm{d}\Omega - \int_{\boldsymbol{x}\in S_0^A}\left(\hat{\boldsymbol{e}}^{\mathrm{AB;I}}_n\times\hat{\boldsymbol{h}}^{\mathrm{BA;I}\circledast}_m + \hat{\boldsymbol{e}}^{\mathrm{BA;I}\circledast}_m\times\hat{\boldsymbol{h}}^{\mathrm{AB;I}}_m\right)\cdot\boldsymbol{\nu}\,\mathrm{d}A \tag{7.31}$$

for all $m = \{1, \dots, N\}$ and $n = \{1, \dots, N\}$, where we have, again, used Eqs. (5.16) and (4.23) and, subsequently, invoked the condition that the resulting identity has to hold for arbitrary exciting electric currents $\hat{I}^{A;\mathrm{T}}_m$ and $\hat{I}^{B;\mathrm{T}}_n$. Taking the limit $s = \delta + \mathrm{i}\omega$ as $\delta \downarrow 0$, we get

$$\hat{Z}^{BA\circledast}_{n,m}(\mathrm{i}\omega) = \frac{1}{2\eta_0}\frac{1}{\lambda_0^2}\int_{\xi\in\Omega}\hat{\boldsymbol{e}}^{\mathrm{BA;I}\circledast}_m(\boldsymbol{\xi},\mathrm{i}\omega)\cdot\hat{\boldsymbol{e}}^{\mathrm{AB;I}}_n(\boldsymbol{\xi},\mathrm{i}\omega)\mathrm{d}\Omega - \int_{\boldsymbol{x}\in S_0^A}\left(\hat{\boldsymbol{e}}^{\mathrm{AB;I}}_n\times\hat{\boldsymbol{h}}^{\mathrm{BA;I}\circledast}_m + \hat{\boldsymbol{e}}^{\mathrm{BA;I}\circledast}_m\times\hat{\boldsymbol{h}}^{\mathrm{AB;I}}_m\right)\cdot\boldsymbol{\nu}\,\mathrm{d}A \tag{7.32}$$

for all $m = \{1, \dots, N\}$ and $n = \{1, \dots, N\}$.

Finally, in the case that conditions (7.22) and (7.23) as well as (7.28) and (7.29) do apply simultaneously, that is, both antenna systems (A) and (B) are loss-free, the reciprocity relation (7.21) can be rewritten as

$$\sum_{m=1}^{M} \left[\hat{V}_m^{A;\mathrm{R}}(s)\hat{I}_m^{A;\mathrm{T}}(-s) - \hat{V}_m^{A;\mathrm{T}}(-s)\hat{I}_m^{A;\mathrm{R}}(s)\right] + \sum_{n=1}^{N} \left[\hat{V}_n^{B;\mathrm{R}}(-s)\hat{I}_n^{B;\mathrm{T}}(s) - \hat{V}_n^{B;\mathrm{T}}(s)\hat{I}_n^{B;\mathrm{R}}(-s)\right] = \left(\eta_0/8\pi^2\right)\int_{\xi\in\Omega} \hat{\boldsymbol{E}}^{\mathrm{AB};\infty}(\boldsymbol{\xi}, s) \cdot \hat{\boldsymbol{E}}^{\mathrm{BA};\infty}(\boldsymbol{\xi}, -s)\mathrm{d}\Omega \tag{7.33}$$

If the receiving antennas are in both states left open-circuited at their terminal ports, the latter reciprocity relation yields

$$\hat{Z}_{m,n}^{AB}(s) + \hat{Z}_{n,m}^{BA}(-s) = -\frac{1}{8\pi^2}\frac{s^2}{c_0^2}\frac{1}{\eta_0}\int_{\xi\in\Omega} \hat{\boldsymbol{e}}_m^{\mathrm{BA;I}}(\boldsymbol{\xi}, -s) \cdot \hat{\boldsymbol{e}}_n^{\mathrm{AB;I}}(\boldsymbol{\xi}, s)\mathrm{d}\Omega \tag{7.34}$$

for all $m = \{1, \dots, M\}$ and $n = \{1, \dots, N\}$. Upon approaching this relation through the real-FD, one finds

$$\hat{Z}_{m,n}^{AB}(\mathrm{i}\omega) + \hat{Z}_{n,m}^{BA\circledast}(\mathrm{i}\omega) = \frac{1}{2\eta_0}\frac{1}{\lambda_0^2}\int_{\xi\in\Omega} \hat{\boldsymbol{e}}_m^{\mathrm{BA;I}\circledast}(\boldsymbol{\xi}, \mathrm{i}\omega) \cdot \hat{\boldsymbol{e}}_n^{\mathrm{AB;I}}(\boldsymbol{\xi}, \mathrm{i}\omega)\mathrm{d}\Omega \tag{7.35}$$

for all $m = \{1, \dots, M\}$ and $n = \{1, \dots, N\}$, where $\circledast$ denotes the complex conjugate. Owing to the property (7.20), we may write

$$\hat{Z}_{m,n}^{AB}(\mathrm{i}\omega) + \hat{Z}_{n,m}^{BA\circledast}(\mathrm{i}\omega) = 2\mathrm{Re}\left[\hat{Z}_{m,n}^{AB}(\mathrm{i}\omega)\right] = 2\mathrm{Re}\left[\hat{Z}_{n,m}^{BA}(\mathrm{i}\omega)\right] \tag{7.36}$$

which in combination with Eq. (7.35) makes possible to calculate the real part of the transfer impedance between two reciprocal and loss-free multiport antenna systems using the electric-current-excited radiated EM far-field constituents.

Exercise

- Let us assume that antenna systems (A) and (B) are reciprocal and loss-free in their EM properties and repeat the reciprocity analysis for the (short-circuit) admittance network parameters.

Hint: The admittance-based description of the remote interaction of two multiport antenna systems in state (BA) is

$$\hat{\boldsymbol{I}}^{A;\mathrm{T}}(s) = \underline{\hat{\boldsymbol{Y}}}^{A;\mathrm{T}}(s) \cdot \hat{\boldsymbol{V}}^{A;\mathrm{T}}(s) + \underline{\hat{\boldsymbol{Y}}}^{AB}(s) \cdot \hat{\boldsymbol{V}}^{B;\mathrm{R}}(s)$$
$$\hat{\boldsymbol{I}}^{B;\mathrm{R}}(s) = \underline{\hat{\boldsymbol{Y}}}^{BA}(s) \cdot \hat{\boldsymbol{V}}^{A;\mathrm{T}}(s) + \underline{\hat{\boldsymbol{Y}}}^{B;\mathrm{L}}(s) \cdot \hat{\boldsymbol{V}}^{B;\mathrm{R}}(s)$$

while for state (AB) we may write

$$\hat{\boldsymbol{I}}^{B;\mathrm{T}}(s) = \underline{\hat{\boldsymbol{Y}}}^{B;\mathrm{T}}(s) \cdot \hat{\boldsymbol{V}}^{B;\mathrm{T}}(s) + \underline{\hat{\boldsymbol{Y}}}^{BA}(s) \cdot \hat{\boldsymbol{V}}^{A;\mathrm{R}}(s)$$
$$\hat{\boldsymbol{I}}^{A;\mathrm{R}}(s) = \underline{\hat{\boldsymbol{Y}}}^{AB}(s) \cdot \hat{\boldsymbol{V}}^{B;\mathrm{T}}(s) + \underline{\hat{\boldsymbol{Y}}}^{A;\mathrm{L}}(s) \cdot \hat{\boldsymbol{V}}^{A;\mathrm{R}}(s)$$

Under the assumption that the antenna systems are reciprocal, the reciprocity relation of the time-convolution type (7.7) directly leads to Eq. (7.19). If the receiving antenna systems in both states are short-circuited, that is, let $\hat{V}_m^{A;\mathrm{R}} = 0$ and $\hat{V}_n^{B;\mathrm{R}} = 0$ for all $m = \{1, \dots, M\}$ and $n = \{1, \dots, N\}$, we get

$$\sum_{m=1}^{M} \hat{V}_m^{A;\mathrm{T}}(s)\hat{I}_m^{A;\mathrm{R}}(s) = \sum_{n=1}^{N} \hat{V}_n^{B;\mathrm{T}}(s)\hat{I}_n^{B;\mathrm{R}}(s)$$

in which the electric currents flowing across the antenna loads are expressed using the relevant admittance parameters, that is,

$$\hat{I}_m^{A;\mathrm{R}}(s) = \sum_{n=1}^{N} \hat{Y}_{m,n}^{AB}(s)\hat{V}_n^{B;\mathrm{T}}(s)|_{\hat{V}_m^{A;\mathrm{R}}=0}$$
$$\hat{I}_n^{B;\mathrm{R}}(s) = \sum_{m=1}^{M} \hat{Y}_{n,m}^{BA}(s)\hat{V}_m^{A;\mathrm{T}}(s)|_{\hat{V}_n^{B;\mathrm{R}}=0}$$

for all $m = \{1, \dots, M\}$ and $n = \{1, \dots, N\}$. Subsequently, upon invoking the condition that the resulting equation has to hold for arbitrary exciting voltages, we get

$$\hat{Y}_{m,n}^{AB}(s) = \hat{Y}_{n,m}^{BA}(s)$$

for all $m = \{1, \dots, M\}$ and $n = \{1, \dots, N\}$, which is equivalent to Eq. (7.20).

Next, under the assumption that the antenna systems are loss-free, the reciprocity relation of the time-correlation type (7.21) directly leads to Eq. (7.33) that for short-circuited receiving antennas has the following form:

$$\sum_{m=1}^{M} \hat{V}_m^{A;T}(-s)\hat{I}_m^{A;R}(s) + \sum_{n=1}^{N} \hat{V}_n^{B;T}(s)\hat{I}_n^{B;R}(-s)$$
$$= -(8\pi^2\eta_0)^{-1} \int_{\xi\in\Omega} \hat{\boldsymbol{H}}^{\mathrm{AB};\infty}(\boldsymbol{\xi}, s) \cdot \hat{\boldsymbol{H}}^{\mathrm{BA};\infty}(\boldsymbol{\xi}, -s)\mathrm{d}\Omega$$

Now, making use of the admittance-based expressions for the electric currents across the antenna loads together with Eq. (5.10), we will end up with:

$$\hat{Y}_{m,n}^{AB}(s) + \hat{Y}_{n,m}^{BA}(-s)$$
$$= \frac{\eta_0}{8\pi^2}\frac{s^2}{c_0^2} \int_{\xi\in\Omega} \hat{\boldsymbol{\ell}}_m^{\mathrm{BA;V}}(\boldsymbol{\xi}, -s) \cdot \hat{\boldsymbol{\ell}}_n^{\mathrm{AB;V}}(\boldsymbol{\xi}, s)\mathrm{d}\Omega$$

for all $m = \{1, \ldots, M\}$ and $n = \{1, \ldots, N\}$. Upon approaching the latter relation through the real-FD, one gets

$$\hat{Y}_{m,n}^{AB}(\mathrm{i}\omega) + \hat{Y}_{n,m}^{BA\circledast}(\mathrm{i}\omega)$$
$$= -\frac{\eta_0}{2}\frac{1}{\lambda_0^2} \int_{\xi\in\Omega} \hat{\boldsymbol{\ell}}_m^{\mathrm{BA;V}\circledast}(\boldsymbol{\xi}, \mathrm{i}\omega) \cdot \hat{\boldsymbol{\ell}}_n^{\mathrm{AB;V}}(\boldsymbol{\xi}, \mathrm{i}\omega)\mathrm{d}\Omega$$

for all $m = \{1, \ldots, M\}$ and $n = \{1, \ldots, N\}$, whose impedance-based equivalent is given in Eq. (7.35).

8

Compensation Theorems for the EM Coupling Between Two Multiport Antennas

The EM field transfer between two antenna systems is inevitably affected by the presence of neighboring objects located in the antenna's (radiative) coupling path. A convenient tool to describe the impact of such an object is found in EM compensation theorems. The concept of compensation is well known in standard electrical network theory where it furnishes an expedient methodology for calculating the effect of a change in the value of a circuit element ([7], Sec. 9.6). The Kirchhoff circuit compensation theorems have been later generalized to its EM form [35], thereby providing an effective instrument to formulate and solve various EM boundary value problems [34]. In this chapter, we shall apply reciprocity theorems of both the time-convolution and time-correlation types to introduce the EM compensation theorems describing the impact of a scattering object on the EM field transfer between two multiport antenna systems.

8.1 Description of the Problem Configuration

In this chapter, we shall interrelate the transmitting–receiving scenarios as shown in Figure 8.1 with those analyzed in Chapter 7. The problem configurations coincide with the ones from Figure 7.1 except for the presence of a scattering object that is placed exterior to the domains occupied by the antenna systems. The scatterer occupies domain $\mathcal{D}$ whose bounding surface is denoted by $\partial\mathcal{D}$. As far as its EM behavior is concerned, the medium of the scatterer is described via the general transverse admittance and the longitudinal impedance, $\underline{\hat{\eta}}^{\mathrm{s}} = \underline{\hat{\eta}}^{\mathrm{s}}(\boldsymbol{x}, s)$ and $\underline{\hat{\zeta}}^{\mathrm{s}} = \underline{\hat{\zeta}}^{\mathrm{s}}(\boldsymbol{x}, s)$, respectively (see Eqs. (1.23) and (1.24)), that include EM penetrable scatterers that are possibly inhomogeneous, anisotropic, and dispersive. In case that the scatterer is perfectly conducting, the explicit-type boundary conditions apply upon approaching $\partial\mathcal{D}$ from its exterior domain (cf. Eqs. (6.10) and (6.11)).

Thanks to the presence of the scatterer, the total fields in the corresponding scenarios are not the same. Accordingly, we shall define the total field difference

Electromagnetic Reciprocity in Antenna Theory, First Edition. Martin Stumpf.

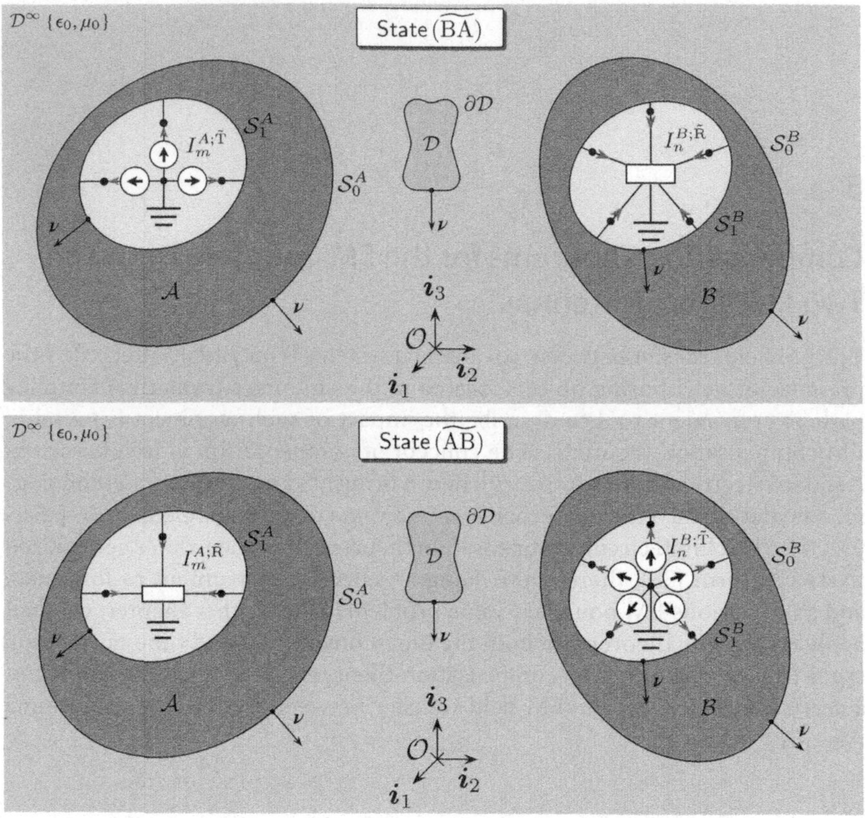

Figure 8.1 State $(\widetilde{\mathrm{BA}})$ with transmitting antenna (A) and receiving antenna (B) and state $(\widetilde{\mathrm{AB}})$ with transmitting antenna (B) and receiving antenna (A). Antenna (A) has $M = 3$ ports and antenna (B) has $N = 5$ ports.

related to state (BA):

$$\{\Delta\hat{\boldsymbol{E}}^{\mathrm{BA}}, \Delta\hat{\boldsymbol{H}}^{\mathrm{BA}}\}(\boldsymbol{x}, s) \triangleq \{\hat{\boldsymbol{E}}^{\widetilde{\mathrm{BA}}} - \hat{\boldsymbol{E}}^{\mathrm{BA}}, \hat{\boldsymbol{H}}^{\widetilde{\mathrm{BA}}} - \hat{\boldsymbol{H}}^{\mathrm{BA}}\}(\boldsymbol{x}, s) \tag{8.1}$$

and, similarly, the total field difference related to state (AB):

$$\{\Delta\hat{\boldsymbol{E}}^{\mathrm{AB}}, \Delta\hat{\boldsymbol{H}}^{\mathrm{AB}}\}(\boldsymbol{x}, s) \triangleq \{\hat{\boldsymbol{E}}^{\widetilde{\mathrm{AB}}} - \hat{\boldsymbol{E}}^{\mathrm{AB}}, \hat{\boldsymbol{H}}^{\widetilde{\mathrm{AB}}} - \hat{\boldsymbol{H}}^{\mathrm{AB}}\}(\boldsymbol{x}, s) \tag{8.2}$$

applying throughout the exterior to $\mathcal{S}_0^A \cup \mathcal{S}_0^B \cup \partial\mathcal{D}$. The corresponding EM field states will be denoted by $(\Delta\mathrm{BA})$ and $(\Delta\mathrm{AB})$. In the following reciprocity analysis, we shall describe the change in the corresponding transfer impedance matrices.

Table 8.1 Application of the reciprocity theorem.

Domain exterior to $\mathcal{S}_0^A \cup \mathcal{S}_0^B \cup \partial\mathcal{D}$		
Time-convolution	State (ΔBA)	State (AB)
Source	0	0
Field	$\{\Delta\hat{\boldsymbol{E}}^{\text{BA}}, \Delta\hat{\boldsymbol{H}}^{\text{BA}}\}$	$\{\hat{\boldsymbol{E}}^{\text{AB}}, \hat{\boldsymbol{H}}^{\text{AB}}\}$
Material	$\{\epsilon_0, \mu_0\}$	$\{\epsilon_0, \mu_0\}$

8.2 Analysis Based on the Reciprocity Theorem of the Time-Convolution Type

In this section, the change in the impedance matrix describing the EM field remote interaction between two multiport antenna systems is analyzed with the aid of the reciprocity theorem of the time-convolution type (see Section 1.4.1). The reciprocity analysis concerning the change in state (BA) and state (AB) are discussed separately in the following sections.

8.2.1 The Change in Scenario (BA)

In order to analyze the effect of the scatterer in state (BA), we start with applying the reciprocity theorem to the unbounded domain exterior to the antenna systems and the scatterer and to (AB) and (ΔBA) EM field states according to Table 8.1. Since the interrelated EM wave fields are causal, the limiting process as described in Section 1.4.3 yields

$$\int_{\boldsymbol{x}\in\mathcal{S}_0^A\cup\mathcal{S}_0^B\cup\partial\mathcal{D}} \left(\Delta\hat{\boldsymbol{E}}^{\text{BA}} \times \hat{\boldsymbol{H}}^{\text{AB}} - \hat{\boldsymbol{E}}^{\text{AB}} \times \Delta\hat{\boldsymbol{H}}^{\text{BA}}\right) \cdot \boldsymbol{\nu}\,\mathrm{d}A = 0 \tag{8.3}$$

If the medium of antenna system (A) is self-adjoint in its EM constitutive properties (see Eqs. (7.8) and (7.9)), the surface integral over its external surface $\mathcal{S}_0^A$ can be directly expressed in terms of the relevant Kirchhoff circuit quantities (see Section 1.5.2), namely

$$\begin{aligned}
&\int_{\boldsymbol{x}\in\mathcal{S}_0^A} \left(\Delta\hat{\boldsymbol{E}}^{\text{BA}} \times \hat{\boldsymbol{H}}^{\text{AB}} - \hat{\boldsymbol{E}}^{\text{AB}} \times \Delta\hat{\boldsymbol{H}}^{\text{BA}}\right) \cdot \boldsymbol{\nu}\,\mathrm{d}A \\
&= \int_{\boldsymbol{x}\in\mathcal{S}_1^A} \left(\Delta\hat{\boldsymbol{E}}^{\text{BA}} \times \hat{\boldsymbol{H}}^{\text{AB}} - \hat{\boldsymbol{E}}^{\text{AB}} \times \Delta\hat{\boldsymbol{H}}^{\text{BA}}\right) \cdot \boldsymbol{\nu}\,\mathrm{d}A \\
&\simeq -\sum_{m=1}^{M} \left[\Delta\hat{V}_m^{A;\text{T}}(s)\hat{I}_m^{A;\text{R}}(s) + \hat{V}_m^{A;\text{R}}(s)\Delta\hat{I}_m^{A;\text{T}}(s)\right]
\end{aligned} \tag{8.4}$$

Table 8.2 Application of the reciprocity theorem.

	Domain $\mathcal{D}$	
Time-convolution	State (BA)	State (AB)
Source	0	0
Field	$\{\hat{\boldsymbol{E}}^{\mathrm{BA}}, \hat{\boldsymbol{H}}^{\mathrm{BA}}\}$	$\{\hat{\boldsymbol{E}}^{\mathrm{AB}}, \hat{\boldsymbol{H}}^{\mathrm{AB}}\}$
Material	$\{\epsilon_0, \mu_0\}$	$\{\epsilon_0, \mu_0\}$

where we have adhered to the orientation of the electric currents with respect to the outer normal vector $\boldsymbol{v}$ along the terminal surface S_1^A (see Figure 8.1). Similar interfacing relation applies to the reciprocal antenna system ($\boldsymbol{B}$), for which we can write

$$\begin{aligned}
&\int_{\boldsymbol{x}\in S_0^B} \left(\Delta\hat{\boldsymbol{E}}^{\mathrm{BA}} \times \hat{\boldsymbol{H}}^{\mathrm{AB}} - \hat{\boldsymbol{E}}^{\mathrm{AB}} \times \Delta\hat{\boldsymbol{H}}^{\mathrm{BA}}\right) \cdot \boldsymbol{v}\,\mathrm{d}A \\
&\quad = \int_{\boldsymbol{x}\in S_1^B} \left(\Delta\hat{\boldsymbol{E}}^{\mathrm{BA}} \times \hat{\boldsymbol{H}}^{\mathrm{AB}} - \hat{\boldsymbol{E}}^{\mathrm{AB}} \times \Delta\hat{\boldsymbol{H}}^{\mathrm{BA}}\right) \cdot \boldsymbol{v}\,\mathrm{d}A \\
&\quad \simeq \sum_{n=1}^{N} \left[\Delta\hat{V}_n^{B;\mathrm{R}}(s)\hat{I}_n^{B;\mathrm{T}}(s) + \hat{V}_n^{B;\mathrm{T}}(s)\Delta\hat{I}_n^{B;\mathrm{R}}(s)\right]
\end{aligned} \tag{8.5}$$

Subsequently, the reciprocity theorem of the time-convolution type is applied to bounded domain $\mathcal{D}$ and to the total fields in states (BA) and (AB) in the absence of scatterer (see Table 8.2). Owing to the self-adjointness of the embedding and thanks to the fact that these states are source-free in $\mathcal{D}$, we get

$$\int_{\boldsymbol{x}\in\partial\mathcal{D}} \left(\hat{\boldsymbol{E}}^{\mathrm{BA}} \times \hat{\boldsymbol{H}}^{\mathrm{AB}} - \hat{\boldsymbol{E}}^{\mathrm{AB}} \times \hat{\boldsymbol{H}}^{\mathrm{BA}}\right) \cdot \boldsymbol{v}\,\mathrm{d}A = 0 \tag{8.6}$$

Finally, upon combining Eqs. (8.3)–(8.6), we get the sought reciprocity relation of the time-convolution type, specifically

$$\begin{aligned}
&\sum_{m=1}^{M} \left[\Delta\hat{V}_m^{A;\mathrm{T}}(s)\hat{I}_m^{A;\mathrm{R}}(s) + \hat{V}_m^{A;\mathrm{R}}(s)\Delta\hat{I}_m^{A;\mathrm{T}}(s)\right] \\
&\quad - \sum_{n=1}^{N} \left[\Delta\hat{V}_n^{B;\mathrm{R}}(s)\hat{I}_n^{B;\mathrm{T}}(s) + \hat{V}_n^{B;\mathrm{T}}(s)\Delta\hat{I}_n^{B;\mathrm{R}}(s)\right] \\
&\quad = \int_{\boldsymbol{x}\in\partial\mathcal{D}} \left(\hat{\boldsymbol{E}}^{\widetilde{\mathrm{BA}}} \times \hat{\boldsymbol{H}}^{\mathrm{AB}} - \hat{\boldsymbol{E}}^{\mathrm{AB}} \times \hat{\boldsymbol{H}}^{\widetilde{\mathrm{BA}}}\right) \cdot \boldsymbol{v}\,\mathrm{d}A
\end{aligned} \tag{8.7}$$

Table 8.3 Application of the reciprocity theorem.

Domain $\mathcal{D}$		
Time-convolution	State $(\widetilde{\text{BA}})$	State (AB)
Source	$\{\hat{\boldsymbol{J}}^{\widetilde{\text{BA}}}, \hat{\boldsymbol{K}}^{\widetilde{\text{BA}}}\}$	0
Field	$\{\hat{\boldsymbol{E}}^{\widetilde{\text{BA}}}, \hat{\boldsymbol{H}}^{\widetilde{\text{BA}}}\}$	$\{\hat{\boldsymbol{E}}^{\text{AB}}, \hat{\boldsymbol{H}}^{\text{AB}}\}$
Material	$\{\epsilon_0, \mu_0\}$	$\{\epsilon_0, \mu_0\}$

from which a number of special cases can be arrived at. For EM penetrable scatterers, the right-hand side of Eq. (8.7) can be expressed in terms of equivalent contrast electric and magnetic current volume densities, that is,

$$\hat{\boldsymbol{J}}^{\widetilde{\text{BA}}}(\boldsymbol{x}, s) = [\, \underline{\hat{\eta}}^{\text{s}}(\boldsymbol{x}, s) - s\epsilon_0 \underline{\boldsymbol{I}}\,] \cdot \hat{\boldsymbol{E}}^{\widetilde{\text{BA}}}(\boldsymbol{x}, s) \tag{8.8}$$

$$\hat{\boldsymbol{K}}^{\widetilde{\text{BA}}}(\boldsymbol{x}, s) = [\, \underline{\hat{\zeta}}^{\text{s}}(\boldsymbol{x}, s) - s\mu_0 \underline{\boldsymbol{I}}\,] \cdot \hat{\boldsymbol{H}}^{\widetilde{\text{BA}}}(\boldsymbol{x}, s) \tag{8.9}$$

for $\boldsymbol{x} \in \mathcal{D}$. Accordingly, upon applying the reciprocity theorem of the time-convolution type in accordance with Table 8.3, we obtain

$$\begin{aligned}
&\int_{\boldsymbol{x}\in\partial\mathcal{D}} \left(\hat{\boldsymbol{E}}^{\widetilde{\text{BA}}} \times \hat{\boldsymbol{H}}^{\text{AB}} - \hat{\boldsymbol{E}}^{\text{AB}} \times \hat{\boldsymbol{H}}^{\widetilde{\text{BA}}}\right) \cdot \boldsymbol{\nu}\,\mathrm{d}A \\
&= \int_{\boldsymbol{x}\in\mathcal{D}} \Big\{ \hat{\boldsymbol{E}}^{\text{AB}} \cdot \left[\underline{\hat{\eta}}^{\text{s}} - s\epsilon_0 \underline{\boldsymbol{I}}\right] \cdot \hat{\boldsymbol{E}}^{\widetilde{\text{BA}}} \\
&\quad - \hat{\boldsymbol{H}}^{\text{AB}} \cdot \left[\underline{\hat{\zeta}}^{\text{s}} - s\mu_0 \underline{\boldsymbol{I}}\right] \cdot \hat{\boldsymbol{H}}^{\widetilde{\text{BA}}} \Big\} \mathrm{d}V
\end{aligned} \tag{8.10}$$

that can be substituted in Eq. (8.7) to get the final reciprocity relation pertaining to EM penetrable scatterers. As an example, let us next closely look at the scenario specified as follows:

1. The transmitting antennas in the both states (BA) and $(\widetilde{\text{BA}})$ are activated by electric currents $\hat{I}_m^{A;\text{T}}$ and $\hat{I}_m^{A;\tilde{\text{T}}}$, respectively, that are identical, that is, $\hat{I}_m^{A;\text{T}} = \hat{I}_m^{A;\tilde{\text{T}}}$ for all $m = \{1, \ldots, M\}$. Accordingly, we let $\Delta\hat{I}_m^{A;\text{T}} = 0$ for all $m = \{1, \ldots, M\}$.
2. The receiving antennas in both states (BA) and $(\widetilde{\text{BA}})$ are open-circuited at all of their accessible ports. Hence, we have $\hat{I}_n^{B;\text{R}} = 0$ and $\hat{I}_n^{B;\tilde{\text{R}}} = 0$ for all $n = \{1, \ldots, N\}$, which implies $\Delta\hat{I}_n^{B;\text{R}} = 0$ for all $n = \{1, \ldots, N\}$.
3. The receiving antenna in state (AB) is open-circuited at all of its terminal ports. Hence, we let $\hat{I}_m^{A;\text{R}} = 0$ for all $m = \{1, \ldots, M\}$.

Table 8.4 Application of the reciprocity theorem.

Domain exterior to $\mathcal{S}_0^A \cup \mathcal{S}_0^B \cup \partial\mathcal{D}$		
Time-convolution	State (ΔAB)	State (BA)
Source	0	0
Field	$\{\Delta\hat{\boldsymbol{E}}^{\text{AB}}, \Delta\hat{\boldsymbol{H}}^{\text{AB}}\}$	$\{\hat{\boldsymbol{E}}^{\text{BA}}, \hat{\boldsymbol{H}}^{\text{BA}}\}$
Material	$\{\epsilon_0, \mu_0\}$	$\{\epsilon_0, \mu_0\}$

Under these conditions, Eq. (8.7) boils down to

$$\sum_{n=1}^{N} \Delta\hat{V}_n^{B;\text{R}}(s)\hat{I}_n^{B;\text{T}}(s) = \sum_{n=1}^{N}\sum_{m=1}^{M} \hat{I}_n^{B;\text{T}}(s)\Delta\hat{Z}_{n,m}^{BA}(s)\hat{I}_m^{A;\text{T}}(s)$$

$$= \int_{\boldsymbol{x}\in\partial\mathcal{D}} \left(\hat{\boldsymbol{E}}^{\text{AB}} \times \hat{\boldsymbol{H}}^{\widetilde{\text{BA}}} - \hat{\boldsymbol{E}}^{\widetilde{\text{BA}}} \times \hat{\boldsymbol{H}}^{\text{AB}}\right) \cdot \boldsymbol{\nu}\,\mathrm{d}A \tag{8.11}$$

where we have used the (open-circuited) impedance parameters in line with Eq. (7.2). Upon invoking the condition that the latter equality has to hold true for arbitrary $\hat{I}_m^{A;\text{T}}$ and $\hat{I}_n^{B;\text{T}}$, we end up with

$$\Delta\hat{Z}_{n,m}^{BA}(s) = \int_{\boldsymbol{x}\in\partial\mathcal{D}} \left(\hat{\boldsymbol{e}}_n^{\text{AB;I}} \times \hat{\boldsymbol{h}}_m^{\widetilde{\text{BA}};\text{I}} - \hat{\boldsymbol{e}}_m^{\widetilde{\text{BA}};\text{I}} \times \hat{\boldsymbol{h}}_n^{\text{AB;I}}\right) \cdot \boldsymbol{\nu}\,\mathrm{d}A \tag{8.12}$$

for all $m = \{1, \dots, M\}$ and $n = \{1, \dots, N\}$ using the (sensing) impulse–electric-current-excited EM wave field constituents defined in Eq. (5.16). Finally, for EM penetrable scatterers, the change in the transfer impedance matrix reads

$$\Delta\hat{Z}_{n,m}^{BA} = -\int_{\boldsymbol{x}\in\mathcal{D}} \Big\{\hat{\boldsymbol{e}}_n^{\text{AB;I}} \cdot \left[\underline{\hat{\boldsymbol{\eta}}}^{\text{s}} - s\epsilon_0\underline{\boldsymbol{I}}\right] \cdot \hat{\boldsymbol{e}}_m^{\widetilde{\text{BA}}}$$

$$- \hat{\boldsymbol{h}}_n^{\text{AB;I}} \cdot \left[\underline{\hat{\boldsymbol{\zeta}}}^{\text{s}} - s\mu_0\underline{\boldsymbol{I}}\right] \cdot \hat{\boldsymbol{h}}_m^{\widetilde{\text{AB}}}\Big\}\mathrm{d}V \tag{8.13}$$

for all $m = \{1, \dots, M\}$ and $n = \{1, \dots, N\}$, where we have applied the equality given in Eq. (8.10).

8.2.2 The Change in Scenario (AB)

In this section, we shall describe the impact of the scatterer on the antenna remote interaction in state (AB) where antenna system (B) operates as a multiport transmitter, while antenna system (A) acts as a multiport loaded scatterer. The reciprocity analysis follows the lines of reasoning from the previous section. Accordingly, we begin with the application of the reciprocity theorem to the exterior domain outside the antenna systems and the scatterer and to states to (BA) and (ΔAB) EM field states (see Table 8.4). In this way, using the conclusions drawn in Section 1.4.3, we get

$$\int_{\boldsymbol{x}\in S_0^A\cup S_0^B\cup\partial D}\left(\Delta\hat{\boldsymbol{E}}^{\mathrm{AB}}\times\hat{\boldsymbol{H}}^{\mathrm{BA}}-\hat{\boldsymbol{E}}^{\mathrm{BA}}\times\Delta\hat{\boldsymbol{H}}^{\mathrm{AB}}\right)\cdot\boldsymbol{\nu}\,\mathrm{d}A=0 \tag{8.14}$$

For antenna systems that are self-adjoint in their EM properties, the surface integral over their externally bounding surface can be directly coupled to the corresponding Kirchhoff circuit quantities, namely,

$$\begin{aligned}&\int_{\boldsymbol{x}\in S_0^A}\left(\Delta\hat{\boldsymbol{E}}^{\mathrm{AB}}\times\hat{\boldsymbol{H}}^{\mathrm{BA}}-\hat{\boldsymbol{E}}^{\mathrm{BA}}\times\Delta\hat{\boldsymbol{H}}^{\mathrm{AB}}\right)\cdot\boldsymbol{\nu}\,\mathrm{d}A\\&=\int_{\boldsymbol{x}\in S_1^A}\left(\Delta\hat{\boldsymbol{E}}^{\mathrm{AB}}\times\hat{\boldsymbol{H}}^{\mathrm{BA}}-\hat{\boldsymbol{E}}^{\mathrm{BA}}\times\Delta\hat{\boldsymbol{H}}^{\mathrm{AB}}\right)\cdot\boldsymbol{\nu}\,\mathrm{d}A\\&\simeq\sum_{m=1}^{M}\left[\Delta\hat{V}_m^{A;\mathrm{R}}(s)\hat{I}_m^{A;\mathrm{T}}(s)+\hat{V}_m^{A;\mathrm{T}}(s)\Delta\hat{I}_m^{A;\mathrm{R}}(s)\right]\end{aligned} \tag{8.15}$$

while the interfacing relation on antenna system (B) reads

$$\begin{aligned}&\int_{\boldsymbol{x}\in S_0^B}\left(\Delta\hat{\boldsymbol{E}}^{\mathrm{AB}}\times\hat{\boldsymbol{H}}^{\mathrm{BA}}-\hat{\boldsymbol{E}}^{\mathrm{BA}}\times\Delta\hat{\boldsymbol{H}}^{\mathrm{AB}}\right)\cdot\boldsymbol{\nu}\,\mathrm{d}A\\&=\int_{\boldsymbol{x}\in S_1^B}\left(\Delta\hat{\boldsymbol{E}}^{\mathrm{AB}}\times\hat{\boldsymbol{H}}^{\mathrm{BA}}-\hat{\boldsymbol{E}}^{\mathrm{BA}}\times\Delta\hat{\boldsymbol{H}}^{\mathrm{AB}}\right)\cdot\boldsymbol{\nu}\,\mathrm{d}A\\&\simeq-\sum_{n=1}^{N}\left[\Delta\hat{V}_n^{B;\mathrm{T}}(s)\hat{I}_n^{B;\mathrm{R}}(s)+\hat{V}_n^{B;\mathrm{R}}(s)\Delta\hat{I}_n^{B;\mathrm{T}}(s)\right]\end{aligned} \tag{8.16}$$

where we have adhered to the conventional orientation of the electric currents on S_1. Combining Eqs. (8.14)–(8.16) with Eq. (8.6) gives the desired reciprocity relation:

$$\begin{aligned}&-\sum_{m=1}^{M}\left[\Delta\hat{V}_m^{A;\mathrm{R}}(s)\hat{I}_m^{A;\mathrm{T}}(s)+\hat{V}_m^{A;\mathrm{T}}(s)\Delta\hat{I}_m^{A;\mathrm{R}}(s)\right]\\&\quad+\sum_{n=1}^{N}\left[\Delta\hat{V}_n^{B;\mathrm{T}}(s)\hat{I}_n^{B;\mathrm{R}}(s)+\hat{V}_n^{B;\mathrm{R}}(s)\Delta\hat{I}_n^{B;\mathrm{T}}(s)\right]\\&=\int_{\boldsymbol{x}\in\partial D}\left(\hat{\boldsymbol{E}}^{\widetilde{\mathrm{AB}}}\times\hat{\boldsymbol{H}}^{\mathrm{BA}}-\hat{\boldsymbol{E}}^{\mathrm{BA}}\times\hat{\boldsymbol{H}}^{\widetilde{\mathrm{AB}}}\right)\cdot\boldsymbol{\nu}\,\mathrm{d}A\end{aligned} \tag{8.17}$$

from which a number of special cases can be arrived at. Similar to the previous section, the right-hand side of Eq. (8.17) can be further expressed in terms of the equivalent contrast electric and magnetic current volume densities of the scattering object, specifically

$$\int_{\boldsymbol{x}\in\partial\mathbb{D}} \left(\hat{\boldsymbol{E}}^{\widetilde{\mathrm{AB}}} \times \hat{\boldsymbol{H}}^{\mathrm{BA}} - \hat{\boldsymbol{E}}^{\mathrm{BA}} \times \hat{\boldsymbol{H}}^{\widetilde{\mathrm{AB}}}\right) \cdot \boldsymbol{\nu}\,\mathrm{d}A$$

$$= \int_{\boldsymbol{x}\in\mathbb{D}} \left\{ \hat{\boldsymbol{E}}^{\mathrm{BA}} \cdot \left[\underline{\hat{\boldsymbol{\eta}}}^{\mathrm{s}} - s\epsilon_0 \underline{\boldsymbol{I}}\right] \cdot \hat{\boldsymbol{E}}^{\widetilde{\mathrm{AB}}} - \hat{\boldsymbol{H}}^{\mathrm{BA}} \cdot \left[\underline{\hat{\boldsymbol{\zeta}}}^{\mathrm{s}} - s\mu_0 \underline{\boldsymbol{I}}\right] \cdot \hat{\boldsymbol{H}}^{\widetilde{\mathrm{AB}}} \right\} \mathrm{d}V \tag{8.18}$$

Substituting the latter equality in Eq. (8.17), we get the final reciprocity relation describing the impact of an EM penetrable scatterer. Finally, let us closely analyze the following situation:

1. The transmitting antennas in both states (AB) and ($\widetilde{\mathrm{AB}}$) are activated by electric currents $\hat{I}_n^{B;\mathrm{T}}$ and $\hat{I}_n^{B;\tilde{\mathrm{T}}}$, respectively, that are identical, that is, $\hat{I}_n^{B;\mathrm{T}} = \hat{I}_n^{B;\tilde{\mathrm{T}}}$ for all $n = \{1, \dots, N\}$. Accordingly, we let $\Delta\hat{I}_n^{B;\mathrm{T}} = 0$ for all $n = \{1, \dots, N\}$.
2. The receiving antennas in both states (AB) and ($\widetilde{\mathrm{AB}}$) are open-circuited at all of their accessible ports. Hence, we have $\hat{I}_m^{A;\mathrm{R}} = 0$ and $\hat{I}_m^{A;\tilde{\mathrm{R}}} = 0$ for all $m = \{1, \dots, M\}$, which implies $\Delta\hat{I}_m^{A;\mathrm{R}} = 0$ for all $m = \{1, \dots, M\}$.
3. The receiving antenna in state (BA) is open-circuited at all of its terminal ports. Hence, we let $\hat{I}_n^{B;\mathrm{R}} = 0$ for all $n = \{1, \dots, N\}$.

Under these conditions, Eq. (8.17) has the following form:

$$\sum_{m=1}^{M} \Delta\hat{V}_m^{A;\mathrm{R}}(s)\hat{I}_m^{A;\mathrm{T}}(s) = \sum_{m=1}^{M}\sum_{n=1}^{N} \hat{I}_m^{A;\mathrm{T}}(s)\Delta\hat{Z}_{m,n}^{AB}(s)\hat{I}_n^{B;\mathrm{T}}(s)$$

$$= \int_{\boldsymbol{x}\in\partial\mathbb{D}} \left(\hat{\boldsymbol{E}}^{\mathrm{BA}} \times \hat{\boldsymbol{H}}^{\widetilde{\mathrm{AB}}} - \hat{\boldsymbol{E}}^{\widetilde{\mathrm{AB}}} \times \hat{\boldsymbol{H}}^{\mathrm{BA}}\right) \cdot \boldsymbol{\nu}\,\mathrm{d}A \tag{8.19}$$

where we have used the (open-circuited) impedance parameters in line with Eq. (7.6). Upon invoking the condition that the latter equality has to hold for arbitrary $\hat{I}_m^{A;\mathrm{T}}$ and $\hat{I}_n^{B;\mathrm{T}}$, we end up with

$$\Delta\hat{Z}_{m,n}^{AB}(s) = \int_{\boldsymbol{x}\in\partial\mathbb{D}} \left(\hat{\boldsymbol{e}}_m^{\mathrm{BA;I}} \times \hat{\boldsymbol{h}}_n^{\widetilde{\mathrm{AB}};\mathrm{I}} - \hat{\boldsymbol{e}}_n^{\widetilde{\mathrm{AB}};\mathrm{I}} \times \hat{\boldsymbol{h}}_m^{\mathrm{BA;I}}\right) \cdot \boldsymbol{\nu}\,\mathrm{d}A \tag{8.20}$$

for all $m = \{1, \dots, M\}$ and $n = \{1, \dots, N\}$. For the sake of completeness, a special form of the latter that applies to EM penetrable scatterers can be written

Table 8.5 Application of the reciprocity theorem.

Domain exterior to $\mathcal{S}_0^A \cup \mathcal{S}_0^B \cup \partial\mathcal{D}$		
Time-correlation	State (ΔBA)	State (AB)
Source	0	0
Field	$\{\Delta\hat{\boldsymbol{E}}^{\mathrm{BA}}, \Delta\hat{\boldsymbol{H}}^{\mathrm{BA}}\}$	$\{\hat{\boldsymbol{E}}^{\mathrm{AB}}, \hat{\boldsymbol{H}}^{\mathrm{AB}}\}$
Material	$\{\epsilon_0, \mu_0\}$	$\{\epsilon_0, \mu_0\}$

as

$$\Delta \hat{Z}^{AB}_{m,n}(s) = -\int_{x\in D} \Big\{ \hat{\boldsymbol{e}}^{\mathrm{BA;I}}_m \cdot \left[\underline{\hat{\boldsymbol{\eta}}}^{\mathrm{s}} - s\epsilon_0\underline{\boldsymbol{I}}\right] \cdot \hat{\boldsymbol{e}}^{\overline{\mathrm{AB}}}_n - \hat{\boldsymbol{h}}^{\mathrm{BA;I}}_m \cdot \left[\underline{\hat{\boldsymbol{\zeta}}}^{\mathrm{s}} - s\mu_0\underline{\boldsymbol{I}}\right] \cdot \hat{\boldsymbol{h}}^{\overline{\mathrm{BA}}}_n \Big\}\mathrm{d}V \tag{8.21}$$

for all $m = \{1, \dots, M\}$ and $n = \{1, \dots, N\}$ in virtue of the equality given in Eq. (8.18). Finally, note that Eqs. (8.12) and (8.20) can be understood as a generalization of the EM compensation theorems that have been previously derived in Eqs. (3) and (4) of Ref. [35] and in Eq. (2.4) of Ref. [34] concerning a 1-port antenna system.

8.3 Analysis Based on the Reciprocity Theorem of the Time-Correlation Type

In this section, the reciprocity theorem of the time-correlation type (see Section 1.4.2) is applied to analyze the remote interaction between two multiport antenna systems. The analysis of the change in the mutual impedance matrices is discussed separately for state (BA) and state (AB).

8.3.1 The Change in Scenario (BA)

To analyze the impact of presence of the scatterer in state (BA), the reciprocity theorem is first applied to the unbounded domain exterior to the antenna systems and the scatterer and to (AB) and (ΔBA) EM field states (see Table 8.5). Since the interrelated EM wave fields are causal, the limiting process as described in Section 1.4.3 yields

$$\int_{\boldsymbol{x}\in S_0^A\cup S_0^B\cup\partial D}\left(\Delta\hat{\boldsymbol{E}}^{\mathrm{BA}\circledast}\times\hat{\boldsymbol{H}}^{\mathrm{AB}}+\hat{\boldsymbol{E}}^{\mathrm{AB}}\times\Delta\hat{\boldsymbol{H}}^{\mathrm{BA}\circledast}\right)\cdot\boldsymbol{\nu}\,\mathrm{d}A$$
$$=\left(\eta_0/8\pi^2\right)\int_{\boldsymbol{\xi}\in\Omega}\hat{\boldsymbol{E}}^{\mathrm{AB};\infty}(\boldsymbol{\xi},s)\cdot\Delta\hat{\boldsymbol{E}}^{\mathrm{BA};\infty}(\boldsymbol{\xi},-s)\mathrm{d}\Omega \tag{8.22}$$

Now, if the medium of antenna system (A) is time-reverse adjoint in its EM properties (see Section 1.5.1), the integral over its bounding surface can be expressed as

$$\int_{\boldsymbol{x}\in S_0^A}\left(\Delta\hat{\boldsymbol{E}}^{\mathrm{BA}\circledast}\times\hat{\boldsymbol{H}}^{\mathrm{AB}}+\hat{\boldsymbol{E}}^{\mathrm{AB}}\times\Delta\hat{\boldsymbol{H}}^{\mathrm{BA}\circledast}\right)\cdot\boldsymbol{\nu}\,\mathrm{d}A$$
$$=\int_{\boldsymbol{x}\in S_1^A}\left(\Delta\hat{\boldsymbol{E}}^{\mathrm{BA}\circledast}\times\hat{\boldsymbol{H}}^{\mathrm{AB}}+\hat{\boldsymbol{E}}^{\mathrm{AB}}\times\Delta\hat{\boldsymbol{H}}^{\mathrm{BA}\circledast}\right)\cdot\boldsymbol{\nu}\,\mathrm{d}A$$
$$\simeq\sum_{m=1}^{M}\left[\hat{V}_m^{A;\mathrm{R}}(s)\Delta\hat{I}_m^{A;\mathrm{T}}(-s)-\Delta\hat{V}_m^{A;\mathrm{T}}(-s)\hat{I}_m^{A;\mathrm{R}}(s)\right] \tag{8.23}$$

where the orientation of the electric currents with respect to $\boldsymbol{\nu}$ along S_1^A has been accounted for. A similar relation applies to antenna system $(\boldsymbol{B})$, namely,

$$\int_{\boldsymbol{x}\in S_0^B}\left(\Delta\hat{\boldsymbol{E}}^{\mathrm{BA}\circledast}\times\hat{\boldsymbol{H}}^{\mathrm{AB}}+\hat{\boldsymbol{E}}^{\mathrm{AB}}\times\Delta\hat{\boldsymbol{H}}^{\mathrm{BA}\circledast}\right)\cdot\boldsymbol{\nu}\,\mathrm{d}A$$
$$=\int_{\boldsymbol{x}\in S_1^B}\left(\Delta\hat{\boldsymbol{E}}^{\mathrm{BA}\circledast}\times\hat{\boldsymbol{H}}^{\mathrm{AB}}+\hat{\boldsymbol{E}}^{\mathrm{AB}}\times\Delta\hat{\boldsymbol{H}}^{\mathrm{BA}\circledast}\right)\cdot\boldsymbol{\nu}\,\mathrm{d}A$$
$$\simeq-\sum_{n=1}^{N}\left[\hat{V}_n^{B;\mathrm{T}}(s)\Delta\hat{I}_n^{B;\mathrm{R}}(-s)-\Delta\hat{V}_n^{B;\mathrm{R}}(-s)\hat{I}_n^{B;\mathrm{T}}(s)\right] \tag{8.24}$$

provided that the medium in $\mathcal{B}$ is time-reverse self-adjoint. In the following step, the reciprocity theorem of the time-correlation type is applied to the domain of the scattering object, $\mathcal{D}$, and to the EM field states (BA) and (AB) (see Figure 7.1 and Table 8.6). Thanks to the time-reverse self-adjointness of the embedding, we get

$$\int_{\boldsymbol{x}\in\partial\mathcal{D}}\left(\hat{\boldsymbol{E}}^{\mathrm{BA}\circledast}\times\hat{\boldsymbol{H}}^{\mathrm{AB}}+\hat{\boldsymbol{E}}^{\mathrm{AB}}\times\hat{\boldsymbol{H}}^{\mathrm{BA}\circledast}\right)\cdot\boldsymbol{\nu}\,\mathrm{d}A=0 \tag{8.25}$$

Table 8.6 Application of the reciprocity theorem.

Domain $\mathcal{D}$		
Time-correlation	State (BA)	State (AB)
Source	0	0
Field	$\{\hat{\boldsymbol{E}}^{\mathrm{BA}}, \hat{\boldsymbol{H}}^{\mathrm{BA}}\}$	$\{\hat{\boldsymbol{E}}^{\mathrm{AB}}, \hat{\boldsymbol{H}}^{\mathrm{AB}}\}$
Material	$\{\epsilon_0, \mu_0\}$	$\{\epsilon_0, \mu_0\}$

Finally, combination of Eqs. (8.22)–(8.25) yields the final reciprocity relation of the time-correlation type:

$$\begin{aligned}
&\sum_{m=1}^{M} \left[\hat{V}_m^{A;\mathrm{R}}(s)\Delta\hat{I}_m^{A;\mathrm{T}}(-s) - \Delta\hat{V}_m^{A;\mathrm{T}}(-s)\hat{I}_m^{A;\mathrm{R}}(s)\right] \\
&+ \sum_{n=1}^{N} \left[\Delta\hat{V}_n^{B;\mathrm{R}}(-s)\hat{I}_n^{B;\mathrm{T}}(s) - \hat{V}_n^{B;\mathrm{T}}(s)\Delta\hat{I}_n^{B;\mathrm{R}}(-s)\right] \\
&= \left(\eta_0/8\pi^2\right) \int_{\boldsymbol{\xi}\in\Omega} \hat{\boldsymbol{E}}^{\mathrm{AB};\infty}(\boldsymbol{\xi}, s) \cdot \Delta\hat{\boldsymbol{E}}^{\mathrm{BA};\infty}(\boldsymbol{\xi}, -s)\mathrm{d}\Omega \\
&- \int_{\boldsymbol{x}\in\partial\mathcal{D}} \left(\hat{\boldsymbol{E}}^{\widetilde{\mathrm{BA}}\circledast} \times \hat{\boldsymbol{H}}^{\mathrm{AB}} + \hat{\boldsymbol{E}}^{\mathrm{AB}} \times \hat{\boldsymbol{H}}^{\widetilde{\mathrm{BA}}\circledast}\right) \cdot \boldsymbol{\nu}\,\mathrm{d}A
\end{aligned} \tag{8.26}$$

which can be viewed as the time-correlation counterpart of Eq. (8.7). The last term on the right-hand side can be further developed for EM-penetrable scattering objects whose presence is accounted for via the equivalent contrast electric and magnetic current volume densities defined in Eqs. (8.8) and (8.9). To this end, the reciprocity theorem is applied in line with Table 8.7 and we end up with

$$\begin{aligned}
&\int_{\boldsymbol{x}\in\partial\mathcal{D}} \left(\hat{\boldsymbol{E}}^{\widetilde{\mathrm{BA}}\circledast} \times \hat{\boldsymbol{H}}^{\mathrm{AB}} + \hat{\boldsymbol{E}}^{\mathrm{AB}} \times \hat{\boldsymbol{H}}^{\widetilde{\mathrm{BA}}\circledast}\right) \cdot \boldsymbol{\nu}\,\mathrm{d}A \\
&\quad = \int_{\boldsymbol{x}\in\mathcal{D}} \Big\{ \hat{\boldsymbol{E}}^{\mathrm{AB}} \cdot \left[\underline{\hat{\boldsymbol{\eta}}}^{\mathrm{s}\circledast} + s\epsilon_0\underline{\boldsymbol{I}}\right] \cdot \hat{\boldsymbol{E}}^{\widetilde{\mathrm{BA}}\circledast} \\
&\qquad + \hat{\boldsymbol{H}}^{\mathrm{AB}} \cdot \left[\underline{\hat{\boldsymbol{\zeta}}}^{\mathrm{s}\circledast} + s\mu_0\underline{\boldsymbol{I}}\right] \cdot \hat{\boldsymbol{H}}^{\widetilde{\mathrm{BA}}\circledast} \Big\}\mathrm{d}V
\end{aligned} \tag{8.27}$$

The latter can be substituted in the reciprocity relation (8.26) to get its special form applying to EM penetrable scattering objects. The conditions considered in Section 8.2.1, that is, $\hat{I}_m^{A;\mathrm{T}} = \hat{I}_m^{A;\tilde{\mathrm{T}}}$, $\hat{I}_n^{B;\mathrm{R}} = \hat{I}_n^{B;\tilde{\mathrm{R}}} = 0$ and $\hat{I}_m^{A;\mathrm{R}} = 0$ for all $m =$

Table 8.7 Application of the reciprocity theorem.

Domain $\mathcal{D}$		
Time-correlation	State $(\widetilde{\mathrm{BA}})$	State (AB)
Source	$\{\hat{\boldsymbol{J}}^{\widetilde{\mathrm{BA}}}, \hat{\boldsymbol{K}}^{\widetilde{\mathrm{BA}}}\}$	0
Field	$\{\hat{\boldsymbol{E}}^{\widetilde{\mathrm{BA}}}, \hat{\boldsymbol{H}}^{\widetilde{\mathrm{BA}}}\}$	$\{\hat{\boldsymbol{E}}^{\mathrm{AB}}, \hat{\boldsymbol{H}}^{\mathrm{AB}}\}$
Material	$\{\epsilon_0, \mu_0\}$	$\{\epsilon_0, \mu_0\}$

$\{1, \ldots, M\}$ and $n = \{1, \ldots, N\}$, allow rewriting Eq. (8.26) as

$$\Delta \hat{Z}^{BA}_{n,m}(-s) = -\frac{1}{8\pi^2}\frac{s^2}{c_0^2}\frac{1}{\eta_0}\int_{\xi\in\Omega} \hat{\boldsymbol{\ell}}^{\mathrm{AB;I}}_n(\xi, s)\cdot\Delta\hat{\boldsymbol{\ell}}^{\mathrm{BA;I}}_m(\xi, -s)\mathrm{d}\Omega - \int_{\boldsymbol{x}\in\partial\mathcal{D}}\left(\hat{\boldsymbol{e}}^{\mathrm{AB;I}}_n \times \hat{\boldsymbol{h}}^{\mathrm{BA;I}\circledast}_m + \hat{\boldsymbol{e}}^{\mathrm{BA;I}\circledast}_m \times \hat{\boldsymbol{h}}^{\mathrm{AB;I}}_n\right)\cdot\boldsymbol{\nu}\,\mathrm{d}A \tag{8.28}$$

upon using Eqs. (4.23), (5.16), and (7.2) and invoking the condition that the equality has to hold for arbitrary $\hat{I}^{A;\mathrm{T}}_m$ and $\hat{I}^{B;\mathrm{T}}_n$ for all $m = \{1, \ldots, M\}$ and $n = \{1, \ldots, N\}$. In the limiting real-FD, the latter expression has the following form (cf. Eq. (6.38)):

$$\Delta \hat{Z}^{BA}_{n,m}(\mathrm{i}\omega) = \frac{1}{2\eta_0}\frac{1}{\lambda_0^2}\int_{\xi\in\Omega} \hat{\boldsymbol{\ell}}^{\mathrm{AB;I}\circledast}_n(\xi, \mathrm{i}\omega)\cdot\Delta\hat{\boldsymbol{\ell}}^{\mathrm{BA;I}}_m(\xi, \mathrm{i}\omega)\mathrm{d}\Omega - \int_{\boldsymbol{x}\in\partial\mathcal{D}}\left(\hat{\boldsymbol{e}}^{\widetilde{\mathrm{BA}};\mathrm{I}}_m \times \hat{\boldsymbol{h}}^{\mathrm{AB;I}\circledast}_n + \hat{\boldsymbol{e}}^{\mathrm{AB;I}\circledast}_n \times \hat{\boldsymbol{h}}^{\widetilde{\mathrm{BA}};\mathrm{I}}_m\right)\cdot\boldsymbol{\nu}\,\mathrm{d}A \tag{8.29}$$

for all $m = \{1, \ldots, M\}$, $n = \{1, \ldots, N\}$, and all $\omega \in \mathbb{R}$ of interest. Finally, employing the equality (8.27) that applies to EM penetrable scattering objects, one finds

$$\Delta \hat{Z}^{BA}_{n,m}(\mathrm{i}\omega) = \frac{1}{2\eta_0}\frac{1}{\lambda_0^2}\int_{\xi\in\Omega} \hat{\boldsymbol{\ell}}^{\mathrm{AB;I}\circledast}_n(\xi, \mathrm{i}\omega)\cdot\Delta\hat{\boldsymbol{\ell}}^{\mathrm{BA;I}}_m(\xi, \mathrm{i}\omega)\mathrm{d}\Omega - \int_{\boldsymbol{x}\in\mathcal{D}}\left\{\hat{\boldsymbol{e}}^{\mathrm{AB;I}\circledast}_n\cdot\left[\underline{\hat{\boldsymbol{\eta}}}^{\mathrm{s}} - \mathrm{i}\omega\epsilon_0\underline{\boldsymbol{I}}\right]\cdot\hat{\boldsymbol{e}}^{\widetilde{\mathrm{BA}};\mathrm{I}}_m + \hat{\boldsymbol{h}}^{\mathrm{AB;I}\circledast}_n\cdot\left[\underline{\hat{\boldsymbol{\zeta}}}^{\mathrm{s}} - \mathrm{i}\omega\mu_0\underline{\boldsymbol{I}}\right]\cdot\hat{\boldsymbol{h}}^{\widetilde{\mathrm{BA}};\mathrm{I}}_m\right\}\mathrm{d}V \tag{8.30}$$

for all $m = \{1, \ldots, M\}$, $n = \{1, \ldots, N\}$, and all $\omega \in \mathbb{R}$.

8.3.2 The Change in Scenario (AB)

The change of the transfer impedance matrix in state (AB) is analyzed with the aid of the time-correlation reciprocity theorem. Similar to the previous

Table 8.8 Application of the reciprocity theorem.

Domain exterior to $\mathcal{S}_0^A \cup \mathcal{S}_0^B \cup \partial\mathcal{D}$		
Time-correlation	State (ΔAB)	State (BA)
Source	0	0
Field	$\{\Delta\hat{\boldsymbol{E}}^{\mathrm{AB}}, \Delta\hat{\boldsymbol{H}}^{\mathrm{AB}}\}$	$\{\hat{\boldsymbol{E}}^{\mathrm{BA}}, \hat{\boldsymbol{H}}^{\mathrm{BA}}\}$
Material	$\{\epsilon_0, \mu_0\}$	$\{\epsilon_0, \mu_0\}$

section, the reciprocity theorem is first applied to the exterior domain and to (ΔAB) and (BA) states (see Table 8.8). Making use of the conclusions drawn in Section 1.4.3, we end up with

$$
\begin{aligned}
&\int_{\boldsymbol{x}\in\mathcal{S}_0^A\cup\mathcal{S}_0^B\cup\partial\mathcal{D}} \left(\Delta\hat{\boldsymbol{E}}^{\mathrm{AB}\circledast}\times\hat{\boldsymbol{H}}^{\mathrm{BA}}+\hat{\boldsymbol{E}}^{\mathrm{BA}}\times\Delta\hat{\boldsymbol{H}}^{\mathrm{AB}\circledast}\right)\cdot\boldsymbol{\nu}\,\mathrm{d}A \\
&= (\eta_0/8\pi^2)\int_{\boldsymbol{\xi}\in\Omega}\hat{\boldsymbol{E}}^{\mathrm{BA};\infty}(\boldsymbol{\xi},s)\cdot\Delta\hat{\boldsymbol{E}}^{\mathrm{AB};\infty}(\boldsymbol{\xi},-s)\mathrm{d}\Omega \qquad (8.31)
\end{aligned}
$$

Under the assumption that the medium in $\mathcal{A}$ is time-reverse self-adjoint in its EM properties, we may write

$$
\begin{aligned}
&\int_{\boldsymbol{x}\in\mathcal{S}_0^A} \left(\Delta\hat{\boldsymbol{E}}^{\mathrm{AB}\circledast}\times\hat{\boldsymbol{H}}^{\mathrm{BA}}+\hat{\boldsymbol{E}}^{\mathrm{BA}}\times\Delta\hat{\boldsymbol{H}}^{\mathrm{AB}\circledast}\right)\cdot\boldsymbol{\nu}\,\mathrm{d}A \\
&\quad= \int_{\boldsymbol{x}\in\mathcal{S}_1^A} \left(\Delta\hat{\boldsymbol{E}}^{\mathrm{AB}\circledast}\times\hat{\boldsymbol{H}}^{\mathrm{BA}}+\hat{\boldsymbol{E}}^{\mathrm{BA}}\times\Delta\hat{\boldsymbol{H}}^{\mathrm{AB}\circledast}\right)\cdot\boldsymbol{\nu}\,\mathrm{d}A \\
&\quad\simeq -\sum_{m=1}^{M}\left[\hat{V}_m^{A;\mathrm{T}}(s)\Delta\hat{I}_m^{A;\mathrm{R}}(-s)-\Delta\hat{V}_m^{A;\mathrm{R}}(-s)\hat{I}_m^{A;\mathrm{T}}(s)\right] \qquad (8.32)
\end{aligned}
$$

while for a time-reverse self-adjoint medium in $\mathcal{B}$, we get

$$
\begin{aligned}
&\int_{\boldsymbol{x}\in\mathcal{S}_0^B} \left(\Delta\hat{\boldsymbol{E}}^{\mathrm{AB}\circledast}\times\hat{\boldsymbol{H}}^{\mathrm{BA}}+\hat{\boldsymbol{E}}^{\mathrm{BA}}\times\Delta\hat{\boldsymbol{H}}^{\mathrm{AB}\circledast}\right)\cdot\boldsymbol{\nu}\,\mathrm{d}A \\
&\quad= \int_{\boldsymbol{x}\in\mathcal{S}_1^B} \left(\Delta\hat{\boldsymbol{E}}^{\mathrm{AB}\circledast}\times\hat{\boldsymbol{H}}^{\mathrm{BA}}+\hat{\boldsymbol{E}}^{\mathrm{BA}}\times\Delta\hat{\boldsymbol{H}}^{\mathrm{AB}\circledast}\right)\cdot\boldsymbol{\nu}\,\mathrm{d}A \\
&\quad\simeq \sum_{n=1}^{N}\left[\hat{V}_n^{B;\mathrm{R}}(s)\Delta\hat{I}_n^{B;\mathrm{T}}(-s)-\Delta\hat{V}_n^{B;\mathrm{T}}(-s)\hat{I}_n^{B;\mathrm{R}}(s)\right] \qquad (8.33)
\end{aligned}
$$

where we have adhered to the orientation of the electric currents with respect to $\boldsymbol{\nu}$ along the antenna terminal surfaces. Combination of Eqs. (8.31)–(8.33) and Eq. (8.25) with the replacement $s \to -s$ yields the final reciprocity relation:

$$\sum_{m=1}^{M} [\Delta\hat{V}_m^{A;R}(-s)\hat{I}_m^{A;T}(s) - \hat{V}_m^{A;T}(s)\Delta\hat{I}_m^{A;R}(-s)]$$
$$+\sum_{n=1}^{N} [\hat{V}_n^{B;R}(s)\Delta\hat{I}_n^{B;T}(-s) - \Delta\hat{V}_n^{B;T}(-s)\hat{I}_n^{B;R}(s)]$$
$$= (\eta_0/8\pi^2)\int_{\xi\in\Omega} \hat{\boldsymbol{E}}^{BA;\infty}(\boldsymbol{\xi},s)\cdot\Delta\hat{\boldsymbol{E}}^{AB;\infty}(\boldsymbol{\xi},-s)\mathrm{d}\Omega$$
$$-\int_{\boldsymbol{x}\in\partial D}\left(\hat{\boldsymbol{E}}^{\widetilde{AB}\circledast}\times\hat{\boldsymbol{H}}^{BA}+\hat{\boldsymbol{E}}^{BA}\times\hat{\boldsymbol{H}}^{\widetilde{AB}\circledast}\right)\cdot\boldsymbol{\nu}\,\mathrm{d}A \tag{8.34}$$

For EM penetrable scatterers, one may further re-write the last integral on the right-hand side as

$$\int_{\boldsymbol{x}\in\partial D}\left(\hat{\boldsymbol{E}}^{\widetilde{AB}\circledast}\times\hat{\boldsymbol{H}}^{BA}+\hat{\boldsymbol{E}}^{BA}\times\hat{\boldsymbol{H}}^{\widetilde{AB}\circledast}\right)\cdot\boldsymbol{\nu}\,\mathrm{d}A$$
$$=\int_{\boldsymbol{x}\in D}\Big\{\hat{\boldsymbol{E}}^{BA}\cdot\left[\underline{\hat{\boldsymbol{\eta}}}^{s\circledast}+s\epsilon_0\underline{\boldsymbol{I}}\right]\cdot\hat{\boldsymbol{E}}^{\widetilde{AB}\circledast}$$
$$+\hat{\boldsymbol{H}}^{BA}\cdot\left[\underline{\hat{\boldsymbol{\zeta}}}^{s\circledast}+s\mu_0\underline{\boldsymbol{I}}\right]\cdot\hat{\boldsymbol{H}}^{\widetilde{AB}\circledast}\Big\}\mathrm{d}V \tag{8.35}$$

upon applying the reciprocity theorem to D and to $(\widetilde{AB})$ and (BA) EM field states. The special case analyzed in Section 8.2.2, that is, $\hat{I}_n^{B;T} = \hat{I}_n^{B;\tilde{T}}$, $\hat{I}_m^{A;R} = \hat{I}_m^{A;\tilde{R}} = 0$, and $\hat{I}_n^{B;R} = 0$ for all $m = \{1,\dots,M\}$ and $n = \{1,\dots,N\}$, simplifies Eq. (8.34) to

$$\Delta\hat{Z}_{m,n}^{AB}(-s) = -\frac{1}{8\pi^2}\frac{s^2}{c_0^2}\frac{1}{\eta_0}\int_{\xi\in\Omega}\hat{\boldsymbol{e}}_m^{BA;I}(\boldsymbol{\xi},s)\cdot\Delta\hat{\boldsymbol{e}}_n^{AB;I}(\boldsymbol{\xi},-s)\mathrm{d}\Omega$$
$$-\int_{\boldsymbol{x}\in\partial D}\left(\hat{\boldsymbol{e}}_m^{BA;I}\times\hat{\boldsymbol{h}}_n^{\widetilde{AB};I\circledast}+\hat{\boldsymbol{e}}_n^{\widetilde{AB};I\circledast}\times\hat{\boldsymbol{h}}_m^{BA;I}\right)\cdot\boldsymbol{\nu}\,\mathrm{d}A \tag{8.36}$$

for all $m = \{1,\dots,M\}$, $n = \{1,\dots,N\}$, where we have, again, used Eq. (7.6). Under the limit $s = \delta + \mathrm{i}\omega$, $\delta \downarrow 0$, the change of the transfer impedance matrix can be evaluated according to (cf. Eq. (6.38))

$$\Delta\hat{Z}_{m,n}^{AB}(\mathrm{i}\omega) = \frac{1}{2\eta_0}\frac{1}{\lambda_0^2}\int_{\xi\in\Omega}\hat{\boldsymbol{e}}_m^{BA;I\circledast}(\boldsymbol{\xi},\mathrm{i}\omega)\cdot\Delta\hat{\boldsymbol{e}}_n^{AB;I}(\boldsymbol{\xi},\mathrm{i}\omega)\mathrm{d}\Omega$$
$$-\int_{\boldsymbol{x}\in\partial D}\left(\hat{\boldsymbol{e}}_n^{\widetilde{AB};I}\times\hat{\boldsymbol{h}}_m^{BA;I\circledast}+\hat{\boldsymbol{e}}_m^{BA;I\circledast}\times\hat{\boldsymbol{h}}_n^{\widetilde{AB};I}\right)\cdot\boldsymbol{\nu}\,\mathrm{d}A \tag{8.37}$$

for all $m = \{1,\dots,M\}$, $n = \{1,\dots,N\}$, and all $\omega \in \mathbb{R}$. Finally, making use of Eq. (8.35), we may write the special form of Eq. (8.37) applying to EM penetrable scatterers, namely,

$$\Delta\hat{Z}^{AB}_{m,n}(\mathrm{i}\omega) = \frac{1}{2\eta_0}\frac{1}{\lambda_0^2}\int_{\xi\in\Omega} \hat{\boldsymbol{\ell}}^{\mathrm{BA;I}\circledast}_n(\boldsymbol{\xi},\mathrm{i}\omega)\cdot\Delta\hat{\boldsymbol{\ell}}^{\mathrm{AB;I}}_m(\boldsymbol{\xi},\mathrm{i}\omega)\mathrm{d}\Omega$$
$$-\int_{\boldsymbol{x}\in D}\Big\{\hat{\boldsymbol{e}}^{\mathrm{BA;I}\circledast}_m\cdot\big[\underline{\hat{\boldsymbol{\eta}}}^{\mathrm{s}}-\mathrm{i}\omega\epsilon_0\underline{\boldsymbol{I}}\big]\cdot\hat{\boldsymbol{e}}^{\widetilde{\mathrm{AB}};\mathrm{I}}_n$$
$$+\hat{\boldsymbol{h}}^{\mathrm{BA;I}\circledast}_m\cdot\big[\underline{\hat{\boldsymbol{\zeta}}}^{\mathrm{s}}-\mathrm{i}\omega\mu_0\underline{\boldsymbol{I}}\big]\cdot\hat{\boldsymbol{h}}^{\widetilde{\mathrm{AB}},\mathrm{I}}_n\Big\}\mathrm{d}V \quad (8.38)$$

for all $m = \{1,\ldots,M\}$, $n = \{1,\ldots,N\}$, and all $\omega \in \mathbb{R}$.

Exercise

- Compensation theorems in EM theory are frequently used in conjunction with the concept of surface impedance (e.g., Ref. [24]). Accordingly, let us analyze the EM field from a vertical electrical dipole over an imperfectly conducting ground and provide an approximate model for the excited horizontal electric field involving only the EM fields for a PEC ground and the surface impedance.

Hint: The EM field components that are parallel (denoted by subscript t) with respect to the conducting ground at $x_3 = 0$ (see Figure 8.2) can be expressed as

$$\hat{\boldsymbol{E}}_t(x_1, x_2, x_3, s) = \hat{\jmath}(s)\boldsymbol{\nabla}_t\partial_3\hat{G}(x_1, x_2, x_3, s)/s\epsilon_0$$
$$\hat{\boldsymbol{H}}_t(x_1, x_2, x_3, s) = \hat{\jmath}(s)[\boldsymbol{\nabla} \times \hat{G}(x_1, x_2, x_3, s)\boldsymbol{i}_3]_t$$

in $x_3 > 0$, where the Green's function satisfies

$$(\boldsymbol{\nabla} \cdot \boldsymbol{\nabla})\hat{G} - (s^2/c_0^2)\hat{G} = 0$$

for $x_3 > 0$ along with the excitation condition:

$$\lim_{x_3 \downarrow h} \partial_3\hat{G} - \lim_{x_3 \uparrow h} \partial_3\hat{G} = -1$$

with $h > 0$ being the source height and the (Leontovich) impedance boundary condition:

$$\lim_{x_3 \downarrow 0} \partial_3\hat{G} = s\epsilon_0\hat{Z}(s) \lim_{x_3 \downarrow 0} \hat{G}$$

where $\hat{Z}(s)$ is the surface impedance describing the reflection properties of the boundary surface. Next, take the advantage of the problem shift invariance and

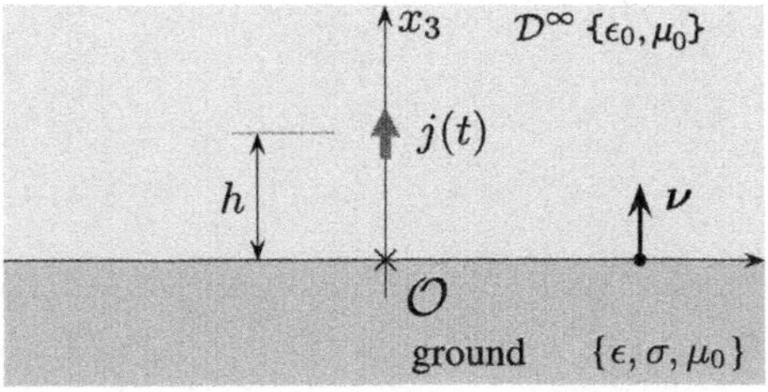

Figure 8.2 Vertical electric dipole above an imperfectly conducting ground.

employ the wave-slowness field representation:

$$\hat{G}(x_1, x_2, x_3, s) = (s/2\pi)^2$$

$$\int_{\alpha_1 \in \mathbb{R}} \mathrm{d}\alpha_1 \int_{\alpha_2 \in \mathbb{R}} \tilde{G}(\alpha_1, \alpha_2, x_3, s) \exp[-\mathrm{i}s(\alpha_1 x_1 + \alpha_2 x_2)] \mathrm{d}\alpha_2$$

which entails $\tilde{\partial}_\kappa = -\mathrm{i}s\alpha_\kappa$ with $\kappa \in \{1, 2\}$ and $\{s \in \mathbb{R}; s > 0\}$. Then show that

$$\tilde{G}(\alpha_1, \alpha_2, x_3, s) = \exp[-s\gamma_0|x_3 - h|]/2s\gamma_0 + \exp[-s\gamma_0(x_3 + h)]/2s\gamma_0$$
$$- \frac{1}{s\gamma_0} \frac{\hat{Z}(s)/\eta_0}{c_0\gamma_0 + \hat{Z}(s)/\eta_0} \exp[-s\gamma_0(x_3 + h)]$$

where the last term on the right-hand side can be interpreted as the correction with respect to the PEC half-space and $\gamma_0 = (c_0^{-2} + \alpha_1^2 + \alpha_2^2)^{1/2}$ with $\mathrm{Re}(\gamma_0) > 0$. Hence, prove that

$$\tilde{\boldsymbol{E}}_\mathrm{t}(\alpha_1, \alpha_2, x_3, s)|_{x_3'=h} = \tilde{\boldsymbol{E}}_\mathrm{t}^{\mathrm{PEC}}(\alpha_1, \alpha_2, x_3, s)|_{x_3'=h}$$
$$+ \hat{Z}(s)\, \boldsymbol{\nu} \times \tilde{\boldsymbol{H}}_\mathrm{t}(\alpha_1, \alpha_2, 0, s)|_{x_3'=x_3+h}$$

where (PEC) denotes the EM field corresponding to the PEC half-space and x_3' is the vertical coordinate of the source dipole. Note that the latter equality does not involve any approximation but still requires the knowledge of the equivalent electric current surface density on the surface of the non perfectly conducting ground, whose calculation is in general a difficult task. An efficient yet reasonably accurate way out of this difficulty is known as the Cooray–Rubinstein formula [36], namely,

$$\hat{\boldsymbol{E}}_\mathrm{t}(x_1, x_2, x_3, s)|_{x_3'=h} \overset{\mathrm{CR}}{\simeq} \hat{\boldsymbol{E}}_\mathrm{t}^{\mathrm{PEC}}(x_1, x_2, x_3, s)|_{x_3'=h}$$
$$+ \hat{Z}(s)\, \boldsymbol{\nu} \times \hat{\boldsymbol{H}}_\mathrm{t}^{\mathrm{PEC}}(x_1, x_2, 0^+, s)|_{x_3'=h}$$

This approximation makes possible to evaluate the EM field excited by the vertical electrical dipole above a non perfectly conducting ground using the corresponding EM wave fields pertaining to the PEC half-space. As the latter fields are known in closed form, the Cooray–Rubinstein formula is in particular suitable for evaluating transient EM fields excited by lightning return strokes. For analytical studies on its validity, we refer the reader to Refs [41,49] and a generalization of the Cooray–Rubinstein formula in the time domain has been introduced in Ref. [44].

9

Compensation Theorems for the EM Scattering of an Antenna System

The control of EM scattering from loaded scatterers finds its wealth applications in radar cross section reduction and antenna characterization [1,28]. Accordingly, we shall next analyze the possibility of controlling the EM antenna scattering of a multiport receiving antenna system by changing its loading impedance matrix. More precisely, the reciprocity theorem of the time-convolution type is applied to demonstrate that the change of antenna EM scattering is proportional to the impulse-excited radiated EM wave field constituents as defined in Section 5.1. Owing to the proportionality to the radiated fields in the corresponding transmitting situation, this part of the total scattered EM wave field can be recognized as the "reradiated" EM wave field. In this chapter, it is shown that the Kirchhoff-type equivalent circuits introduced in Chapter 5 cannot be in general used to calculate the power scattered by an antenna system but merely to find its re-radiated EM power. Although the latter quantity has no direct physical significance on its own, it is useful for gaining an insight into the adequacy of the Kirchhoff-type equivalent networks in analyzing the EM antenna scattering.

9.1 Description of the Problem Configuration

In this section, we shall describe the impact of a change in the antenna load on EM scattering of an N-port receiving antenna system. To this end, two receiving scenarios differing in the antenna load only are analyzed with the aid of the reciprocity theorem of the time-convolution type (see Figure 9.1). The receiving states are denoted by (R) and ($\tilde{\mathrm{R}}$) and the corresponding load impedance matrices are $\hat{Z}^{\mathrm{L}}_{m,n}$ and $\hat{Z}^{\tilde{\mathrm{L}}}_{m,n}$ for $m = \{1, \ldots, N\}$ and $n = \{1, \ldots, N\}$, respectively. The receiving antenna systems are in both scenarios irradiated either by the uniform plane wave (see Eqs. (3.1) and (3.2)) or by the incident EM wave field generated by external sources. Consequently, the change in EM scattering behavior is described using the scattered field difference, namely,

$$\{\Delta\hat{\boldsymbol{E}}^{\mathrm{s}}, \Delta\hat{\boldsymbol{H}}^{\mathrm{s}}\}(\boldsymbol{x}, s) \triangleq \{\hat{\boldsymbol{E}}^{\tilde{\mathrm{s}}} - \hat{\boldsymbol{E}}^{\mathrm{s}}, \hat{\boldsymbol{H}}^{\tilde{\mathrm{s}}} - \hat{\boldsymbol{H}}^{\mathrm{s}}\}(\boldsymbol{x}, s) \qquad \text{(6.1 revisited)}$$

Electromagnetic Reciprocity in Antenna Theory, First Edition. Martin Stumpf.

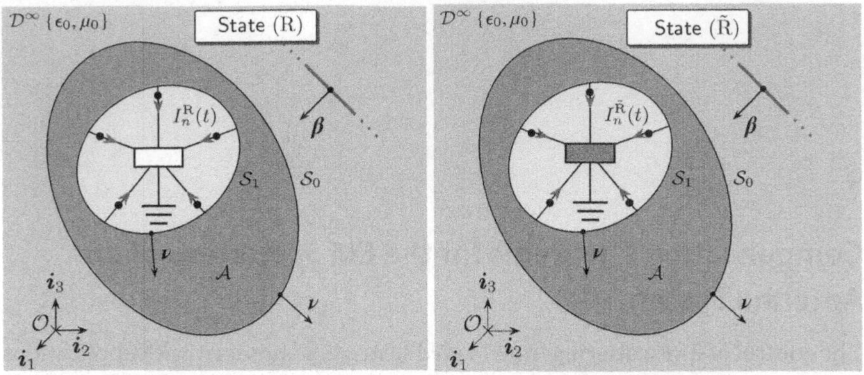

Figure 9.1 Receiving situations differing from each other in the antenna load.

that is, owing to the one and the same incident field in (R) and ($\tilde{\mathrm{R}}$) states, identical to the change in the total EM wave field in the problem configuration, specifically

$$\{\Delta\hat{\boldsymbol{E}}^{\mathrm{R}}, \Delta\hat{\boldsymbol{H}}^{\mathrm{R}}\}(\boldsymbol{x}, s) = \{\Delta\hat{\boldsymbol{E}}^{\mathrm{s}}, \Delta\hat{\boldsymbol{H}}^{\mathrm{s}}\}(\boldsymbol{x}, s) \tag{9.1}$$

The change in the total EM wave fields manifests itself through the change of the Kirchhoff circuit quantities associated with the antenna load (see Eqs. (4.12) and (4.13)). Hence, we define

$$\{\Delta\hat{I}_n^{\mathrm{R}}, \Delta\hat{V}_n^{\mathrm{R}}\}(s) \triangleq \{\hat{I}_n^{\tilde{\mathrm{R}}} - \hat{I}_n^{\mathrm{R}}, \hat{V}_n^{\tilde{\mathrm{R}}} - \hat{V}_n^{\mathrm{R}}\} \tag{9.2}$$

for all $n = \{1, \dots, N\}$. Again, the corresponding EM field states will be further denoted by (Δs) and (ΔR). In order to express the change of the scattered EM wave field in the antenna embedding, the relevant EM field states will be interrelated with the testing field state (B) that is defined as a receiving situation in which the antenna system is irradiated by (fundamental) testing electric or magnetic current volume densities:

$$\{\hat{\boldsymbol{J}}^{\mathrm{B}}, \hat{\boldsymbol{K}}^{\mathrm{B}}\}(\boldsymbol{x}, s) = \{\hat{\imath}^{\mathrm{B}}, \hat{\kappa}^{\mathrm{B}}\}(s)\, \boldsymbol{l}^{\mathrm{B}}\delta(\boldsymbol{x} - \boldsymbol{x}^S) \tag{9.3}$$

with $\boldsymbol{x}^S \in \mathcal{D}^\infty$, where $\hat{\imath}^{\mathrm{B}}$ and $\hat{\kappa}^{\mathrm{B}}$ describe the source amplitudes and $\boldsymbol{l}^{\mathrm{B}}$ is a unit vector describing the source orientation (see Fig. 9.2). Finally, the corresponding transmitting scenario (T) is shown in Figure 4.1.

9.2 Reciprocity Analysis

In the first step of the analysis, we apply the reciprocity theorem of the time-convolution type to the domain exterior to the antenna system and to (ΔR)

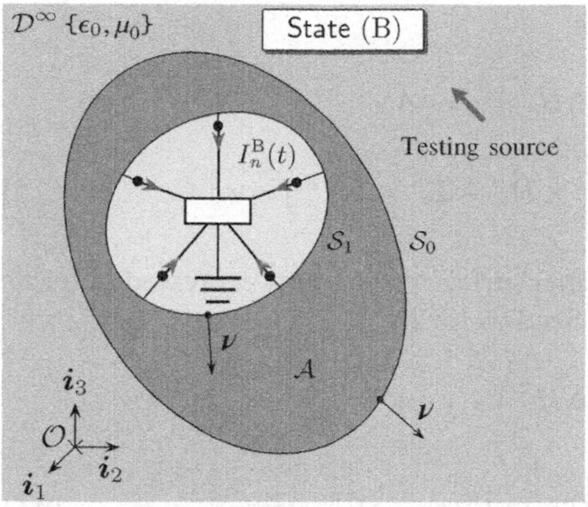

Figure 9.2 Receiving antenna system activated by the testing source.

and (B) states according to Table 9.1. Using the generic form of the reciprocity theorem (1.30) and the conclusions of Section 1.4.3, we first get

$$\int_{\boldsymbol{x}\in\mathcal{S}_0}\left(\Delta\hat{\boldsymbol{E}}^{\mathrm{R}}\times\hat{\boldsymbol{H}}^{\mathrm{B}}-\hat{\boldsymbol{E}}^{\mathrm{B}}\times\Delta\hat{\boldsymbol{H}}^{\mathrm{R}}\right)\cdot\boldsymbol{\nu}\,\mathrm{d}A$$
$$=\int_{\boldsymbol{x}\in\mathcal{D}^{\infty}}\left(\hat{\boldsymbol{J}}^{\mathrm{B}}\cdot\Delta\hat{\boldsymbol{E}}^{\mathrm{R}}-\hat{\boldsymbol{K}}^{\mathrm{B}}\cdot\Delta\hat{\boldsymbol{H}}^{\mathrm{R}}\right)\mathrm{d}V \qquad (9.4)$$

where we have taken into account the orientation of the outer unit vector along $\mathcal{S}_0$. In the following step, the reciprocity theorem is applied to the antenna domain $\mathcal{A}$ and to (ΔR) and (B) states, again. If the medium in $\mathcal{A}$ is reciprocal in its EM properties, the surface field interactions over the antenna bounding

Table 9.1 Application of the reciprocity theorem.

Domain $\mathcal{D}^{\infty}$		
Time-convolution	State (ΔR)	State (B)
Source	0	$\{\hat{\boldsymbol{J}}^{\mathrm{B}}, \hat{\boldsymbol{K}}^{\mathrm{B}}\}$
Field	$\{\Delta\hat{\boldsymbol{E}}^{\mathrm{R}}, \Delta\hat{\boldsymbol{H}}^{\mathrm{R}}\}$	$\{\hat{\boldsymbol{E}}^{\mathrm{B}}, \hat{\boldsymbol{H}}^{\mathrm{B}}\}$
Material	$\{\epsilon_0, \mu_0\}$	$\{\epsilon_0, \mu_0\}$

surfaces are equal, and we obtain

$$\int_{\boldsymbol{x}\in\mathcal{S}_0} \left(\Delta\hat{\boldsymbol{E}}^{\mathrm{R}} \times \hat{\boldsymbol{H}}^{\mathrm{B}} - \hat{\boldsymbol{E}}^{\mathrm{B}} \times \Delta\hat{\boldsymbol{H}}^{\mathrm{R}}\right) \cdot \boldsymbol{v}\,\mathrm{d}A = \int_{\boldsymbol{x}\in\mathcal{S}_1} \left(\Delta\hat{\boldsymbol{E}}^{\mathrm{R}} \times \hat{\boldsymbol{H}}^{\mathrm{B}} - \hat{\boldsymbol{E}}^{\mathrm{B}} \times \Delta\hat{\boldsymbol{H}}^{\mathrm{R}}\right) \cdot \boldsymbol{v}\,\mathrm{d}A \tag{9.5}$$

The surface integral over the terminal surface can be next expressed in terms of Kirchhoff circuit quantities (see Section 1.5.2):

$$\int_{\boldsymbol{x}\in\mathcal{S}_1} \left(\Delta\hat{\boldsymbol{E}}^{\mathrm{R}} \times \hat{\boldsymbol{H}}^{\mathrm{B}} - \hat{\boldsymbol{E}}^{\mathrm{B}} \times \Delta\hat{\boldsymbol{H}}^{\mathrm{R}}\right) \cdot \boldsymbol{v}\,\mathrm{d}A = \sum_{n=1}^{N} \left[\hat{V}_n^{\mathrm{B}}(s)\Delta\hat{I}_n^{\mathrm{R}}(s) - \Delta\hat{V}_n^{\mathrm{R}}(s)\hat{I}_n^{\mathrm{B}}(s)\right] \tag{9.6}$$

where the sign is taken in accordance with the convention that the electric currents in the receiving situations are oriented *into* the load. Combination of Eqs. (9.4)–(9.6) with (9.1) and (9.3) then yields

$$\hat{\imath}^{\mathrm{B}}(s)\,\boldsymbol{l}^{\mathrm{B}} \cdot \Delta\hat{\boldsymbol{E}}^{\mathrm{s}}(\boldsymbol{x}^S, s) - \hat{\kappa}^{\mathrm{B}}(s)\,\boldsymbol{l}^{\mathrm{B}} \cdot \Delta\hat{\boldsymbol{H}}^{\mathrm{s}}(\boldsymbol{x}^S, s) = \sum_{n=1}^{N} \left[\hat{V}_n^{\mathrm{B}}(s)\Delta\hat{I}_n^{\mathrm{R}}(s) - \Delta\hat{V}_n^{\mathrm{R}}(s)\hat{I}_n^{\mathrm{B}}(s)\right] \tag{9.7}$$

for $\boldsymbol{x}^S \in \mathcal{D}^\infty$, which relates the change of the scattered EM wave field with the testing field state (B).

The following task to explore is the (reciprocity) relation between the transmitting situation (T) and the testing field state (B). To this end, the reciprocity theorem is applied to the unbounded domain $\mathcal{D}^\infty$ according to Table 9.2, which

Table 9.2 Application of the reciprocity theorem.

Domain $\mathcal{D}^\infty$		
Time-convolution	State (T)	State (B)
Source	0	$\{\hat{\boldsymbol{J}}^{\mathrm{B}}, \hat{\boldsymbol{K}}^{\mathrm{B}}\}$
Field	$\{\hat{\boldsymbol{E}}^{\mathrm{T}}, \hat{\boldsymbol{H}}^{\mathrm{T}}\}$	$\{\hat{\boldsymbol{E}}^{\mathrm{B}}, \hat{\boldsymbol{H}}^{\mathrm{B}}\}$
Material	$\{\epsilon_0, \mu_0\}$	$\{\epsilon_0, \mu_0\}$

leads to

$$\int_{\boldsymbol{x}\in\mathcal{S}_0}\left(\hat{\boldsymbol{E}}^{\mathrm{T}}\times\hat{\boldsymbol{H}}^{\mathrm{B}}-\hat{\boldsymbol{E}}^{\mathrm{B}}\times\hat{\boldsymbol{H}}^{\mathrm{T}}\right)\cdot\boldsymbol{\nu}\,\mathrm{d}A$$
$$=\int_{\boldsymbol{x}\in\mathcal{D}^{\infty}}\left(\hat{\boldsymbol{J}}^{\mathrm{B}}\cdot\hat{\boldsymbol{E}}^{\mathrm{T}}-\hat{\boldsymbol{K}}^{\mathrm{B}}\cdot\hat{\boldsymbol{H}}^{\mathrm{T}}\right)\mathrm{d}V \tag{9.8}$$

where, again, we have used the facts that both states are causal and the embedding is self-adjoint in its EM behavior. The next step, again, consists of applying the reciprocity theorem to the bounded domain $\mathcal{A}$ and of interfacing the EM wave field quantities on the terminal surface with the relevant Kirchhoff circuit quantities. In this way, we end up with

$$\int_{\boldsymbol{x}\in\mathcal{S}_0}\left(\hat{\boldsymbol{E}}^{\mathrm{T}}\times\hat{\boldsymbol{H}}^{\mathrm{B}}-\hat{\boldsymbol{E}}^{\mathrm{B}}\times\hat{\boldsymbol{H}}^{\mathrm{T}}\right)\cdot\boldsymbol{\nu}\,\mathrm{d}A$$
$$=-\sum_{n=1}^{N}\left[\hat{V}_n^{\mathrm{B}}(s)\hat{I}_n^{\mathrm{T}}(s)+\hat{V}_n^{\mathrm{T}}(s)\hat{I}_n^{\mathrm{B}}(s)\right] \tag{9.9}$$

provided that the receiving antenna system is reciprocal. Hence, combining Eqs. (9.8) and (9.9) with Eq. (9.3), we find the sought reciprocity relation between the testing (B) and transmitting (T) states, specifically:

$$\hat{\imath}^{\mathrm{B}}(s)\,\boldsymbol{l}^{\mathrm{B}}\cdot\hat{\boldsymbol{E}}^{\mathrm{T}}(\boldsymbol{x}^{S},s)-\hat{\kappa}^{\mathrm{B}}(s)\,\boldsymbol{l}^{\mathrm{B}}\cdot\hat{\boldsymbol{H}}^{\mathrm{T}}(\boldsymbol{x}^{S},s)$$
$$=-\sum_{n=1}^{N}\left[\hat{V}_n^{\mathrm{B}}(s)\hat{I}_n^{\mathrm{T}}(s)+\hat{V}_n^{\mathrm{T}}(s)\hat{I}_n^{\mathrm{B}}(s)\right] \tag{9.10}$$

for $\boldsymbol{x}^{S}\subset\mathcal{D}^{\infty}$. The final relations (9.7) and (9.10) can be next combined to relate the change of the antenna EM scattering with the electric-current and voltage-impulse-excited radiated EM wave fields. The two cases shall be next discussed separately.

9.2.1 Compensation Theorems in Terms of Electric Current-excited Sensing EM Fields

To express the change in the antenna EM scattering using the electric-current impulse-excited sensing EM wave fields, the reciprocity relation (9.10) is first re-written using Eq. (5.16), namely

$$\hat{\imath}^{\mathrm{B}}(s)\,\boldsymbol{l}^{\mathrm{B}}\cdot\hat{\boldsymbol{e}}_n^{\mathrm{T;I}}(\boldsymbol{x}^{S},s)\,\hat{I}_n^{\mathrm{T}}(s)-\hat{\kappa}^{\mathrm{B}}(s)\,\boldsymbol{l}^{\mathrm{B}}\cdot\hat{\boldsymbol{h}}_n^{\mathrm{T;I}}(\boldsymbol{x}^{S},s)\,\hat{I}_n^{\mathrm{T}}(s)$$
$$=-\hat{V}_n^{\mathrm{B}}(s)\hat{I}_n^{\mathrm{T}}(s)-\hat{V}_n^{\mathrm{T}}(s)\hat{I}_n^{\mathrm{B}}(s) \tag{9.11}$$

which in view of problem linearity applies to all $n=\{1,\dots,N\}$. Considering the electric-current-excited testing wave fields with the open-circuited receiving antenna system in state (B), that is, $\hat{\kappa}^{\mathrm{B}}(s) = 0$ and $\hat{I}_n^{\mathrm{B}}(s) = 0$ for all $n = \{1, \ldots, N\}$, the combination of Eqs. (9.7) and (9.11) leads to

$$\Delta\hat{\boldsymbol{E}}^{\mathrm{s}}(\boldsymbol{x}^S, s) = -\sum_{n=1}^{N} \hat{\boldsymbol{e}}_n^{\mathrm{T;I}}(\boldsymbol{x}^S, s)\Delta\hat{I}_n^{\mathrm{R}}(s) \tag{9.12}$$

for $\boldsymbol{x}^S \in \mathcal{D}^\infty$, where we have invoked the condition that the resulting condition has to hold for arbitrary $\hat{\imath}^{\mathrm{B}}(s)\,\boldsymbol{l}^{\mathrm{B}}$. Similarly, the nonzero magnetic current testing source with the open-circuited antenna system in situation (B) results in the change of the scattered magnetic field:

$$\Delta\hat{\boldsymbol{H}}^{\mathrm{s}}(\boldsymbol{x}^S, s) = -\sum_{n=1}^{N} \hat{\boldsymbol{h}}_n^{\mathrm{T;I}}(\boldsymbol{x}^S, s)\Delta\hat{I}_n^{\mathrm{R}}(s) \tag{9.13}$$

for $\boldsymbol{x}^S \in \mathcal{D}^\infty$. Finally, note that the compensation theorems can also be expressed using the corresponding far-field EM wave quantities. For instance, the far-field counterpart of Eq. (9.12) has the following form:

$$\Delta\hat{\boldsymbol{E}}^{\mathrm{s};\infty}(\boldsymbol{\xi}, s) = -s\mu_0 \sum_{n=1}^{N} \hat{\boldsymbol{e}}_n^{\mathrm{T;I}}(\boldsymbol{\xi}, s)\Delta\hat{I}_n^{\mathrm{R}}(s) \tag{9.14}$$

as $|\boldsymbol{x}^S| \to \infty$, for $\boldsymbol{\xi} \in \Omega = \{\boldsymbol{\xi} \cdot \boldsymbol{\xi} = 1\}$ on the unit sphere (see Eqs. (3.4), (4.10), and (4.23)) and $\Delta\hat{\boldsymbol{H}}^{\mathrm{s};\infty} = \eta_0\,\boldsymbol{\xi} \times \Delta\hat{\boldsymbol{E}}^{\mathrm{s};\infty}$ follows.

9.2.2 Compensation Theorems in Terms of Voltage-Excited Sensing EM Fields

The change in the antenna scattering can also be described with the aid of the voltage-impulse-excited sensing EM wave fields defined in Eq. (5.18). To this end, the latter is used in Eq. (9.10) and we get

$$\begin{aligned}\hat{\imath}^{\mathrm{B}}(s)\,\boldsymbol{l}^{\mathrm{B}} \cdot \hat{\boldsymbol{e}}_n^{\mathrm{T;V}}(\boldsymbol{x}^S, s)\,\hat{V}_n^{\mathrm{T}}(s) - \hat{\kappa}^{\mathrm{B}}(s)\,\boldsymbol{l}^{\mathrm{B}} \cdot \hat{\boldsymbol{h}}_n^{\mathrm{T;V}}(\boldsymbol{x}^S, s)\,\hat{V}_n^{\mathrm{T}}(s)\\ = -\hat{V}_n^{\mathrm{B}}(s)\hat{I}_n^{\mathrm{T}}(s) - \hat{V}_n^{\mathrm{T}}(s)\hat{I}_n^{\mathrm{B}}(s)\end{aligned} \tag{9.15}$$

for all $n = \{1, \ldots, N\}$. Assuming now that the receiving antenna system in state (B) is short-circuited at all of its accessible ports, that is, $\hat{V}_n^{\mathrm{B}}(s) = 0$ for all $n = \{1, \ldots, N\}$, the change in the scattered electric field strength can be written as

$$\Delta\hat{\boldsymbol{E}}^{\mathrm{s}}(\boldsymbol{x}^S, s) = \sum_{n=1}^{N} \hat{\boldsymbol{e}}_n^{\mathrm{T;V}}(\boldsymbol{x}^S, s)\Delta\hat{V}_n^{\mathrm{R}}(s) \tag{9.16}$$

for $\boldsymbol{x}^S \in \mathcal{D}^\infty$, where we have taken $\hat{\kappa}^{\mathrm{B}}(s) = 0$ and invoked the condition that the resulting relation has to hold for arbitrary $\hat{\imath}^{\mathrm{B}}(s)\,\boldsymbol{l}^{\mathrm{B}}$. Following the similar lines of reasoning, the change of the scattered magnetic field strength reads

$$\Delta \hat{\boldsymbol{H}}^{\mathrm{s}}(\boldsymbol{x}^S, s) = \sum_{n=1}^{N} \hat{\boldsymbol{h}}_n^{\mathrm{T;V}}(\boldsymbol{x}^S, s)\Delta \hat{V}_n^{\mathrm{R}}(s) \tag{9.17}$$

for $\boldsymbol{x}^S \in \mathcal{D}^\infty$. Again, as $|\boldsymbol{x}^S| \to \infty$, the concept of the far-field applies and we may write

$$\Delta \hat{\boldsymbol{H}}^{\mathrm{s};\infty}(\boldsymbol{\xi}, s) = s\epsilon_0 \sum_{n=1}^{N} \hat{\boldsymbol{e}}_n^{\mathrm{T;V}}(\boldsymbol{\xi}, s)\Delta \hat{V}_n^{\mathrm{R}}(s) \tag{9.18}$$

for example, for all $\boldsymbol{\xi} \in \Omega = \{\boldsymbol{\xi} \cdot \boldsymbol{\xi} = 1\}$, where we have used Eq. (5.10). Finally, the electric-type far-field amplitude follows from $\eta_0 \Delta \hat{\boldsymbol{E}}^{\mathrm{s};\infty} = \Delta \hat{\boldsymbol{H}}^{\mathrm{s};\infty} \times \boldsymbol{\xi}$.

9.2.3 Power Reciprocity Expressions

The compensation theorems derived in Sections 9.2.1 and 9.2.2 are next applied to find power-related antenna reciprocity relations that are useful for understanding the limitations of the Kirchhoff-type equivalent networks of a multiport receiving antenna system. To this end, we start with the antenna power–reciprocity relation associated with the transmitting situation (cf. Eq. (1.46)):

$$\begin{aligned}
&\int_{\boldsymbol{x}\in\mathcal{S}_0} \left(\hat{\boldsymbol{E}}^{\mathrm{T}} \times \hat{\boldsymbol{H}}^{\mathrm{T}\circledast} + \hat{\boldsymbol{E}}^{\mathrm{T}\circledast} \times \hat{\boldsymbol{H}}^{\mathrm{T}}\right) \cdot \boldsymbol{\nu}\,\mathrm{d}A \\
&= \int_{\boldsymbol{x}\in\mathcal{S}_1} \left(\hat{\boldsymbol{E}}^{\mathrm{T}} \times \hat{\boldsymbol{H}}^{\mathrm{T}\circledast} + \hat{\boldsymbol{E}}^{\mathrm{T}\circledast} \times \hat{\boldsymbol{H}}^{\mathrm{T}}\right) \cdot \boldsymbol{\nu}\,\mathrm{d}A \\
&\quad - \int_{\boldsymbol{x}\in\mathcal{A}} \Big\{\hat{\boldsymbol{H}}^{\mathrm{T}} \cdot \left[\underline{\hat{\zeta}}^{\circledast} + \underline{\hat{\zeta}}^{T}\right] \cdot \hat{\boldsymbol{H}}^{\mathrm{T}\circledast} \\
&\qquad + \hat{\boldsymbol{E}}^{\mathrm{T}} \cdot \left[\underline{\hat{\eta}}^{\circledast} + \underline{\hat{\eta}}^{T}\right] \cdot \hat{\boldsymbol{E}}^{\mathrm{T}\circledast}\Big\}\mathrm{d}V
\end{aligned} \tag{9.19}$$

that is found upon applying the reciprocity theorem of the time-correlation type to the antenna domain $\mathcal{A}$ and to one and the same transmitting state (T). The interaction surface integral over the antenna bounding surface can be next expressed using the radiated far-field amplitudes (see Eqs. (1.45) and (4.10)):

$$\begin{aligned}
&\int_{\boldsymbol{x}\in\mathcal{S}_0} \left(\hat{\boldsymbol{E}}^{\mathrm{T}} \times \hat{\boldsymbol{H}}^{\mathrm{T}\circledast} + \hat{\boldsymbol{E}}^{\mathrm{T}\circledast} \times \hat{\boldsymbol{H}}^{\mathrm{T}}\right) \cdot \boldsymbol{\nu}\,\mathrm{d}A \\
&= \left(\eta_0/8\pi^2\right) \int_{\boldsymbol{\xi}\in\Omega} \hat{\boldsymbol{E}}^{\mathrm{T};\infty}(\boldsymbol{\xi}, s) \cdot \hat{\boldsymbol{E}}^{\mathrm{T};\infty}(\boldsymbol{\xi}, -s)\mathrm{d}\Omega
\end{aligned} \tag{9.20}$$

where we have used the time-reverse self-adjointness of the antenna embedding together with the property of causality of the transmitted EM wave field. Furthermore, the surface integration on the antenna terminal surface is expressed in terms of the corresponding Kirchhoff circuit quantities:

$$\int_{\boldsymbol{x}\in S_1} \left(\hat{\boldsymbol{E}}^{\mathrm{T}} \times \hat{\boldsymbol{H}}^{\mathrm{T}\circledast} + \hat{\boldsymbol{E}}^{\mathrm{T}\circledast} \times \hat{\boldsymbol{H}}^{\mathrm{T}} \right) \cdot \boldsymbol{\nu}\,\mathrm{d}A$$

$$\simeq \sum_{n=1}^{N} [\hat{V}_n^{\mathrm{T}}(s)\hat{I}_n^{\mathrm{T}}(-s) + \hat{V}_n^{\mathrm{T}}(-s)\hat{I}_n^{\mathrm{T}}(s)] \tag{9.21}$$

where we have adhered to the conventional orientation of the electric currents in the transmitting situation. For loss-free reciprocal antenna systems, the combination of Eqs. (9.19)–(9.21) with Eqs. (4.2) and (4.23) yields

$$-(s^2/c_0^2)(32\pi^2\eta_0)^{-1}\int_{\boldsymbol{\xi}\in\Omega} \hat{\boldsymbol{\ell}}_m^{\mathrm{T;I}}(\boldsymbol{\xi}, s) \cdot \hat{\boldsymbol{\ell}}_n^{\mathrm{T;I}}(\boldsymbol{\xi}, -s)\mathrm{d}\Omega$$

$$= \tfrac{1}{4}[\hat{Z}_{m,n}^{\mathrm{T}}(s) + \hat{Z}_{n,m}^{\mathrm{T}}(-s)] \tag{9.22}$$

for all $m = \{1, \ldots, N\}$ and $n = \{1, \ldots, N\}$. Subsequently, the power-related quantity corresponding to the electric-current impulse-excited compensation theorem (9.14) has the following form:

$$\frac{\eta_0}{32\pi^2}\int_{\boldsymbol{\xi}\in\Omega} \Delta\hat{\boldsymbol{E}}^{\mathrm{S};\infty}(\boldsymbol{\xi}, s) \cdot \Delta\hat{\boldsymbol{E}}^{\mathrm{S};\infty}(\boldsymbol{\xi}, -s)\mathrm{d}\Omega = -(s^2/c_0^2)(32\pi^2\eta_0)^{-1}$$

$$\sum_{m=1}^{N} \Delta\hat{I}_m^{\mathrm{R}}(s) \sum_{n=1}^{N} \Delta\hat{I}_n^{\mathrm{R}}(-s) \int_{\boldsymbol{\xi}\in\Omega} \hat{\boldsymbol{\ell}}_m^{\mathrm{T;I}}(\boldsymbol{\xi}, s) \cdot \hat{\boldsymbol{\ell}}_n^{\mathrm{T;I}}(\boldsymbol{\xi}, -s)\mathrm{d}\Omega \tag{9.23}$$

which in combination with Eq. (9.14) gives the desired power reciprocity relation between the change of the EM scattered field and the relevant Kirchhoff quantities, namely

$$\tfrac{1}{4}\sum_{m=1}^{N}\sum_{n=1}^{N}[\Delta\hat{I}_m^{\mathrm{R}}(s)\hat{Z}_{m,n}^{\mathrm{T}}(s)\Delta\hat{I}_n^{\mathrm{R}}(-s) + \Delta\hat{I}_n^{\mathrm{R}}(-s)\hat{Z}_{n,m}^{\mathrm{T}}(-s)\Delta\hat{I}_m^{\mathrm{R}}(s)]$$

$$= (\eta_0/32\pi^2)\int_{\boldsymbol{\xi}\in\Omega} \Delta\hat{\boldsymbol{E}}^{\mathrm{S};\infty}(\boldsymbol{\xi}, s) \cdot \Delta\hat{\boldsymbol{E}}^{\mathrm{S};\infty}(\boldsymbol{\xi}, -s)\mathrm{d}\Omega \tag{9.24}$$

Now, if the receiving antenna system in state (R) is *open-circuited* at all of its accessible ports, Eq. (9.24) can be, in the limit $\{s = \delta + i\omega, \delta \downarrow 0, \omega \in \mathbb{R}\}$, interpreted as the (time-averaged) EM power dissipated internally in the relevant *Thévenin* equivalent network. Similarly, making use of the voltage-excited effective length defined in Eq. (5.10) with Eq. (4.3), one may arrive at

$$\frac{1}{4}\sum_{m=1}^{N}\sum_{n=1}^{N}[\Delta\hat{V}_m^{\mathrm{R}}(s)\hat{Y}_{m,n}^{\mathrm{T}}(s)\Delta\hat{V}_n^{\mathrm{R}}(-s)+\Delta\hat{V}_n^{\mathrm{R}}(-s)\hat{Y}_{n,m}^{\mathrm{T}}(-s)\Delta\hat{V}_m^{\mathrm{R}}(s)]$$
$$=(32\pi^2\eta_0)^{-1}\int_{\xi\in\Omega}\Delta\hat{\boldsymbol{H}}^{\mathrm{s};\infty}(\boldsymbol{\xi},s)\cdot\Delta\hat{\boldsymbol{H}}^{\mathrm{s};\infty}(\boldsymbol{\xi},-s)\mathrm{d}\Omega \qquad (9.25)$$

which for the *short-circuited* receiving antenna system in the receiving state (R) yields the expression for the internal power dissipated in the relevant *Norton* equivalent network. In conclusion, it has been just demonstrated that the internal power dissipated in the Kirchhoff-type equivalent circuits can be related to the (re-radiated) power quantity associated with the change of the scattered EM wave field. The latter, in general, is not equal to the total EM power scattered by the antenna system. An important exception in this respect are small dipole antennas that do not show structural EM scattering. For more details on the subject, we refer the reader to Refs [40,42] and to the references therein.

Exercises

- Give special forms of the derived compensation theorems concerning a 1-port (a) open-circuited, (b) short-circuited, and (c) matched receiving antenna in state (R).

Hint: If the receiving antenna in state (R) is *open-circuited*, we have $\hat{I}^{\mathrm{R}}(s) = 0$ and Eq. (9.12) with $N = 1$ leads to (cf. Ref. [10], Eq. (10))

$$\begin{aligned}\Delta\hat{\boldsymbol{E}}^{\mathrm{s}}(\boldsymbol{x}^S, s)|_{\hat{Z}^{\mathrm{L}}\to\infty} &= \hat{\boldsymbol{E}}^{\tilde{\mathrm{s}}}(\boldsymbol{x}^S, s) - \hat{\boldsymbol{E}}^{\mathrm{s}}(\boldsymbol{x}^S, s)|_{\hat{Z}^{\mathrm{L}}\to\infty}\\ &= -\,\hat{\boldsymbol{e}}^{\mathrm{T;I}}(\boldsymbol{x}^S, s)\hat{I}^{\tilde{\mathrm{R}}}(s)\\ &= -\,\hat{\boldsymbol{e}}^{\mathrm{T;I}}(\boldsymbol{x}^S, s)\hat{I}^{\mathrm{G}}(s)\hat{Y}^{\tilde{\mathrm{L}}}(s)/\left[\hat{Y}^{\mathrm{T}}(s) + \hat{Y}^{\tilde{\mathrm{L}}}(s)\right]\end{aligned}$$

Equivalently, making use of the voltage-excited compensation theorem (9.16), we may express the change as

$$\begin{aligned}\Delta\hat{\boldsymbol{E}}^{\mathrm{s}}(\boldsymbol{x}^S, s)|_{\hat{Z}^{\mathrm{L}}\to\infty} &= \hat{\boldsymbol{e}}^{\mathrm{T;V}}(\boldsymbol{x}^S, s)\Delta\hat{V}^{\mathrm{R}}(s)\\ &= -\,\hat{\boldsymbol{e}}^{\mathrm{T;V}}(\boldsymbol{x}^S, s)\hat{V}^{\mathrm{G}}(s)\hat{Y}^{\tilde{\mathrm{L}}}(s)/\left[\hat{Y}^{\mathrm{T}}(s) + \hat{Y}^{\tilde{\mathrm{L}}}(s)\right]\end{aligned}$$

which reveals that the sensing impulse-excited EM wave fields are interrelated via the radiation impedance, that is, $\hat{\boldsymbol{e}}^{\mathrm{T;I}} = \hat{\boldsymbol{e}}^{\mathrm{T;V}}\hat{Z}^{\mathrm{T}}$. Similarly, if the receiving antenna in state (R) is *short-circuited*, we have $\hat{V}^{\mathrm{R}}(s) = 0$ and Eq. (9.16) with $N = 1$ leads to

$$\begin{aligned}\Delta\hat{\boldsymbol{E}}^{\mathrm{s}}(\boldsymbol{x}^S, s)|_{\hat{Z}^{\mathrm{L}}\downarrow 0} &= \hat{\boldsymbol{E}}^{\tilde{\mathrm{s}}}(\boldsymbol{x}^S, s) - \hat{\boldsymbol{E}}^{\mathrm{s}}(\boldsymbol{x}^S, s)|_{\hat{Z}^{\mathrm{L}}\downarrow 0}\\ &= \hat{\boldsymbol{e}}^{\mathrm{T;V}}(\boldsymbol{x}^S, s)\hat{V}^{\tilde{\mathrm{R}}}(s)\\ &= \hat{\boldsymbol{e}}^{\mathrm{T;V}}(\boldsymbol{x}^S, s)\hat{V}^{\mathrm{G}}(s)\hat{Z}^{\tilde{\mathrm{L}}}(s)/\left[\hat{Z}^{\mathrm{T}}(s) + \hat{Z}^{\tilde{\mathrm{L}}}(s)\right]\end{aligned}$$

Equivalently, the electric-current-excited compensation theorem (9.12) yields (cf. Ref. [10] Eq. (9))

$$\begin{aligned}\Delta\hat{\boldsymbol{E}}^{\mathrm{s}}(\boldsymbol{x}^S, s)|_{\hat{Z}^{\mathrm{L}}\downarrow 0} &= -\hat{\boldsymbol{e}}^{\mathrm{T;I}}(\boldsymbol{x}^S, s)\Delta\hat{I}^{\mathrm{R}}(s)\\ &= \hat{\boldsymbol{e}}^{\mathrm{T;I}}(\boldsymbol{x}^S, s)\hat{I}^{\mathrm{G}}(s)\hat{Z}^{\tilde{\mathrm{L}}}(s)/\left[\hat{Z}^{\mathrm{T}}(s) + \hat{Z}^{\tilde{\mathrm{L}}}(s)\right]\end{aligned}$$

Finally, if the receiving antenna is *matched* to its transmitting state, the electric-current-excited compensation theorem (9.12) leads to

$$\begin{aligned}\Delta\hat{\boldsymbol{E}}^{\mathrm{s}}(\boldsymbol{x}^S, s)|_{\hat{Z}^{\mathrm{L}}\to\hat{Z}^{\mathrm{T}}} &= \hat{\boldsymbol{E}}^{\tilde{\mathrm{s}}}(\boldsymbol{x}^S, s) - \hat{\boldsymbol{E}}^{\mathrm{s}}(\boldsymbol{x}^S, s)|_{\hat{Z}^{\mathrm{L}}\to\hat{Z}^{\mathrm{T}}}\\ &= -\,\hat{\boldsymbol{e}}^{\mathrm{T;I}}(\boldsymbol{x}^S, s)\Delta\hat{I}^{\mathrm{R}}(s) = -\tfrac{1}{2}\hat{\boldsymbol{e}}^{\mathrm{T;I}}(\boldsymbol{x}^S, s)\hat{I}^{\mathrm{G}}(s)\hat{\Gamma}(s)\end{aligned}$$

while its voltage-excited counterpart (9.16) gives

$$\begin{aligned}\Delta\hat{\boldsymbol{E}}^{\mathrm{s}}(\boldsymbol{x}^S,s)|_{\hat{Z}^{\mathrm{L}}\to\hat{Z}^{\mathrm{T}}} &= \hat{\boldsymbol{e}}^{\mathrm{T;V}}(\boldsymbol{x}^S,s)\Delta\hat{V}^{\mathrm{R}}(s)\\ &= -\frac{1}{2}\hat{\boldsymbol{e}}^{\mathrm{T;V}}(\boldsymbol{x}^S,s)\hat{V}^{\mathrm{G}}(s)\hat{\Gamma}(s)\end{aligned}$$

where we have introduced the reflection coefficient

$$\begin{aligned}\hat{\Gamma}(s) &= \left[\hat{Z}^{\mathrm{T}}(s)-\hat{Z}^{\tilde{\mathrm{L}}}(s)\right]/\left[\hat{Z}^{\mathrm{T}}(s)+\hat{Z}^{\tilde{\mathrm{L}}}(s)\right]\\ &= \left[\hat{Y}^{\tilde{\mathrm{L}}}(s)-\hat{Y}^{\mathrm{T}}(s)\right]/\left[\hat{Y}^{\mathrm{T}}(s)+\hat{Y}^{\tilde{\mathrm{L}}}(s)\right]\end{aligned}$$

Antenna scattering compensation theorems in terms of the reflection coefficient have been previously discussed by Hansen [25], for instance.

- Apply the electric-current impulse-excited compensation theorem (9.12) and the forward-scattering theorem (3.12) to the case of open-circuited, electric-current-excited small dipole antennas.

Hint: At first, it is observed that the small dipole antennas share the properties

$$\begin{aligned}\boldsymbol{\alpha}\cdot\hat{\boldsymbol{\ell}}^{\mathrm{T;I}}(\boldsymbol{\beta},-s) &= \boldsymbol{\alpha}\cdot\hat{\boldsymbol{\ell}}^{\mathrm{T;I}}(-\boldsymbol{\beta},s)\\ \boldsymbol{\alpha}\cdot\hat{\boldsymbol{\ell}}^{\mathrm{T;I}}(\boldsymbol{\beta},s) &= \boldsymbol{\alpha}\cdot\hat{\boldsymbol{\ell}}^{\mathrm{T;I}}(-\boldsymbol{\beta},-s)\end{aligned}$$

where $\boldsymbol{\alpha}$ is a unit vector in the direction of polarization and $\boldsymbol{\beta}$ is a unit vector in the direction of propagation of the incident EM plane wave. Indeed, for a short wire carrying a uniform electric current, we may write

$$\boldsymbol{\alpha}\cdot\hat{\boldsymbol{\ell}}^{\mathrm{T;I}}(\pm\boldsymbol{\beta},s) = -\boldsymbol{\alpha}\cdot\boldsymbol{\ell}$$

irrespective of s, where $\boldsymbol{\ell}$ is the vectorial length of the wire, while for a small loop carrying a uniform electric current, we have

$$\begin{aligned}\boldsymbol{\alpha}\cdot\hat{\boldsymbol{\ell}}^{\mathrm{T;I}}(-\boldsymbol{\beta},s) &= sc_0^{-1}\boldsymbol{A}\cdot(\boldsymbol{\beta}\times\boldsymbol{\alpha}) = \boldsymbol{\alpha}\cdot\hat{\boldsymbol{\ell}}^{\mathrm{T;I}}(\boldsymbol{\beta},-s)\\ \boldsymbol{\alpha}\cdot\hat{\boldsymbol{\ell}}^{\mathrm{T;I}}(\boldsymbol{\beta},s) &= sc_0^{-1}\boldsymbol{A}\cdot(\boldsymbol{\alpha}\times\boldsymbol{\beta}) = \boldsymbol{\alpha}\cdot\hat{\boldsymbol{\ell}}^{\mathrm{T;I}}(-\boldsymbol{\beta},-s)\end{aligned}$$

where $\boldsymbol{A}$ is the vectorial area of the loop. The compensation theorem Eq. (9.12) is then used to express the change of the co-polarized EM scattered field in the forward direction and we get

$$\begin{aligned}(s\mu_0)^{-1}\boldsymbol{\alpha}\cdot\Delta\hat{\boldsymbol{E}}^{\mathrm{S};\infty}(\boldsymbol{\beta},-s)|_{\hat{Z}^{\mathrm{L}}\to\infty} &= \boldsymbol{\alpha}\cdot\hat{\boldsymbol{\ell}}^{\mathrm{T;I}}(-\boldsymbol{\beta},s)\hat{I}^{\tilde{\mathrm{R}}}(-s)\\ -(s\mu_0)^{-1}\boldsymbol{\alpha}\cdot\Delta\hat{\boldsymbol{E}}^{\mathrm{S};\infty}(\boldsymbol{\beta},s)|_{\hat{Z}^{\mathrm{L}}\to\infty} &= \boldsymbol{\alpha}\cdot\hat{\boldsymbol{\ell}}^{\mathrm{T;I}}(-\boldsymbol{\beta},-s)\hat{I}^{\tilde{\mathrm{R}}}(s)\end{aligned}$$

Making use of the expression for the equivalent Thévenin voltage-source strength (5.9), we arrive at (cf. Eq. (3.12))

$$\frac{1}{4}\left[\hat{e}^{\mathrm{i}}(s)\,\boldsymbol{\alpha}\cdot\hat{\boldsymbol{E}}^{\mathrm{s};\infty}(\boldsymbol{\beta},-s)-\hat{e}^{\mathrm{i}}(-s)\,\boldsymbol{\alpha}\cdot\hat{\boldsymbol{E}}^{\mathrm{s};\infty}(\boldsymbol{\beta},s)\right]/s\mu_0$$
$$=\frac{1}{4}\hat{I}^{\tilde{\mathrm{R}}}(s)\left[\hat{Z}^{\mathrm{T}}(s)+\hat{Z}^{\mathrm{T}}(-s)+\hat{Z}^{\mathrm{L}}(s)+\hat{Z}^{\mathrm{L}}(-s)\right]\hat{I}^{\tilde{\mathrm{R}}}(-s)$$

where we have neglected the (structural) EM scattering of an open-circuited small antenna. The right-hand side of the last relation can be in the limit $\{s=\delta+\mathrm{i}\omega,\delta\downarrow 0,\omega\in\mathbb{R}\}$ interpreted as the power dissipated in the internal and load impedances of Thévenin's equivalent circuit. In accordance with the general form of the forward-scattering theorem (3.12), the internal power can be, for such a special class of antennas, associated with the scattered power, while the power dissipated in the antenna load corresponds to the absorbed power.

Appendix A

Lerch's Uniqueness Theorem

The one-sided Laplace transformation was defined as (cf. (1.9))

$$\hat{f}(s) = \int_{t=0}^{\infty} \exp(-st) f(t) \mathrm{d}t \tag{A.1}$$

with $\{s \in \mathbb{C}; \mathrm{Re}(s) > s_0\}$ being its transformation parameter. Interpreting Eq. (1.9) as an integral equation, an important question as to its uniqueness arises. This question has been settled by a Czech mathematician Matyáš Lerch (see Refs [31]; [52], Ch. II; and [20], Sec. 5) such that we may state:

Theorem A.1 Two original functions $g(t)$ and $q(t)$, whose image functions are identical, that is, $\hat{g}(s) \equiv \hat{q}(s)$, differ at most by a null function $n(t)$ for which

$$\int_{\tau=0}^{t} n(\tau) \mathrm{d}\tau = 0 \tag{A.2}$$

for all $t \geq 0$ and hence $\hat{n}(s) = 0$.

The question of uniqueness can be hence reduced to demonstrating that the corresponding homogeneous integral equation, $\hat{f}(s) = 0$, does not posses a nonzero solution.

A.1 Problem of Moments

To prove the theorem, we start by proving the lemma for a problem of moments.

Lemma A.1 Let $\psi(x)$ be a continuous function in $\{0 \leq x \leq 1\}$ and let

$$\int_{x=0}^{1} x^m \psi(x) \mathrm{d}x = 0 \quad \text{for all } m = \{0, 1, \ldots\}. \tag{A.3}$$

Then $\psi(x) \equiv 0$ in $\{0 \leq x \leq 1\}$.

The lemma is next proved by contradiction along the lines presented in Appendix I of Ref. [4]. Hence, if $\psi(x)$ is not identically zero in $\{0 \leq x \leq 1\}$, there must be an interval $\{a \leq x \leq b\}$, with $0 < a < b < 1$, where $\psi(x)$ is either positive or negative. Without loss of generality, we shall next assume the former case:

$$\psi(x) = 0 \quad \text{for } \{0 \leq x < a\} \cup \{b < x \leq 1\} \tag{A.4}$$

$$\psi(x) > 0 \quad \text{for } \{a \leq x \leq b\} \tag{A.5}$$

Let us next consider a function that is linear in x:

$$h(x) = (x/a)\mathrm{H}(x) - [(x - x_0)/a]\mathrm{H}(x - x_0) - [(x - x_0)/(1 - b)]\mathrm{H}(x - x_0) + [(x - 1)/(1 - b)]\mathrm{H}(x - 1) \tag{A.6}$$

with $x_0 = a/(1 - b + a)$ and $\mathrm{H}(x)$ denotes the Heaviside unit step function. Then, in virtue of (A.3), we should get

$$\int_{x=0}^{1} h^m(x)\psi(x)\mathrm{d}x = 0 \quad \text{for all } m = \{0, 1, \dots\} \tag{A.7}$$

since $h^m(x)$ is a polynomial in x. Next observe that

$$0 < h(x) < 1 \quad \text{for } \{0 \leq x < a\} \cup \{b < x \leq 1\} \tag{A.8}$$

$$h(x) \geq 1 \quad \text{for } \{a \leq x \leq b\} \tag{A.9}$$

which implies that $h^m(x)$ can be made as small as we wish in $\{0 \leq x < a\} \cup \{b < x \leq 1\}$ and as large as we wish in $\{a < x < b\}$ for the increasing positive integer m (see Fig. A.1). Consequently, our assumption (A.4)–(A.5) leads to

$$\int_{x=0}^{1} h^m(x)\psi(x)\mathrm{d}x > 0 \quad \text{for all } m = \{0, 1, \dots\} \tag{A.10}$$

This is, however, in contradiction to (A.7) and hence to (A.3). It thus follows that $\psi(x) \equiv 0$ in $\{0 \leq x \leq 1\}$ and the lemma is proved.

A.2 Proof of Lerch's Theorem

The proof of uniqueness of Eq. (A.1) is carried out along the Lerch sequence $\mathcal{L} = \{s \in \mathbb{R}; s = s_0 + nh, h > 0, n = 1, 2, \dots\}$, namely,

$$\hat{f}(s_0 + nh) = \int_{t=0}^{\infty} \exp(-nht)\exp(-s_0 t) f(t)\mathrm{d}t = 0 \tag{A.11}$$

In order to relax the requirement that $f(t)$ is a continuous function of time, we do not apply the lemma to Eq. (A.11) but instead, we begin with the integration

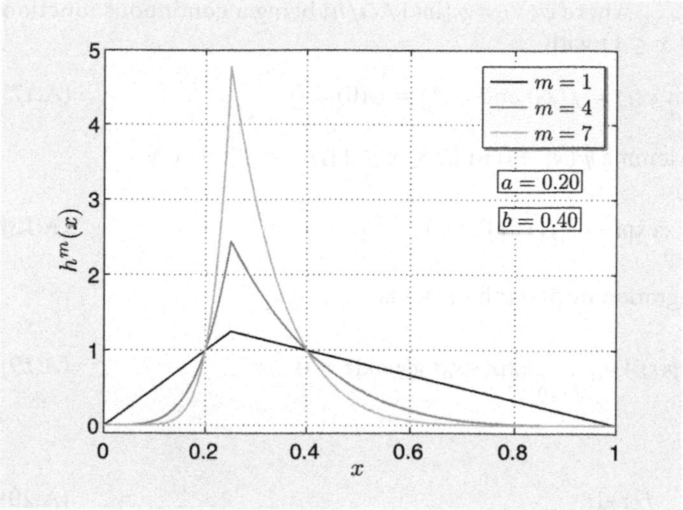

Figure A.1 The polynomial functions.

by parts and get

$$\exp(-nht)\varphi(t)\Big|_{t=0}^{\infty} + nh\int_{t=0}^{\infty} \exp(-nht)\varphi(t)\mathrm{d}t = 0 \tag{A.12}$$

where

$$\varphi(t) = \int_{\tau=0}^{t} \exp(-s_0\tau)f(\tau)\mathrm{d}\tau \tag{A.13}$$

which implies that

$$\int_{t=0}^{\infty} \exp(-nht)\varphi(t)\mathrm{d}t = 0 \tag{A.14}$$

To cast Eq. (A.14) into the form of (A.3), we substitute

$$x = \exp(-ht) \tag{A.15}$$

and get

$$h^{-1}\int_{\lambda-0}^{1} x^{n-1}\psi(x)\mathrm{d}x = 0 \tag{A.16}$$

for all $n = \{1, 2, \dots\}$, where $\psi(x) = \varphi[\ln(1/x)/h]$ being a continuous function throughout $\{0 \leq x \leq 1\}$ with

$$\psi(0) = \lim_{t\to\infty} \varphi(t) = \hat{f}(s_0) \text{ and } \psi(1) = \varphi(0) = 0 \tag{A.17}$$

Hence, from the lemma $\psi(x) \equiv 0$ in $\{0 \leq x \leq 1\}$, or, equivalently

$$\varphi(t) = \int_{\tau=0}^{t} \exp(-s_0\tau) f(\tau) \mathrm{d}\tau \equiv 0 \tag{A.18}$$

for all $t \geq 0$. Integration by parts then yields

$$\exp(-s_0 t)\phi(t) + s_0 \int_{\tau=0}^{t} \exp(-s_0\tau)\phi(\tau) \mathrm{d}\tau = 0 \tag{A.19}$$

where

$$\phi(t) = \int_{\tau=0}^{t} f(\tau) \mathrm{d}\tau \tag{A.20}$$

is a continuous function for all $t \geq 0$. Consequently, we may differentiate Eq. (A.19) with respect to t and find

$$\exp(-s_0 t)[\mathrm{d}\phi(t)/\mathrm{d}t] = 0 \tag{A.21}$$

which with the help of (A.20) yields

$$\frac{\mathrm{d}}{\mathrm{d}t} \int_{\tau=0}^{t} f(\tau) \mathrm{d}\tau = 0 \tag{A.22}$$

Owing to the fact that $\lim_{t\downarrow 0} \phi(t) = 0$, we have

$$\int_{\tau=0}^{t} f(\tau) \mathrm{d}\tau = 0 \tag{A.23}$$

for all $t \geq 0$, which implies that $f(t)$ is a mere null function. Since the null functions can be for physical and technical applications ignored, we have just proved that the Laplace transformation (A.1) of a causal function $f(t)$ is unique for our purposes.

References

1 J. Appel-Hansen. Accurate determination of gain and radiation patterns by radar cross-section measurements. *IEEE Transactions on Antennas and Propagation*, 27(5):640–646, 1979.

2 S. Ballantine. The Lorentz reciprocity theorem for electric waves. *Proceedings of the Institute of Radio Engineers*, 16:513–518, 1928.

3 N. N. Bojarski. Generalized reaction principles and reciprocity theorems for the wave equations, and the relationship between the time-advanced and time-retarded fields. *The Journal of the Acoustical Society of America*, 74(1):281–285, 1983.

4 H. S. Carslaw and J. C. Jaeger. *Operational Methods in Applied Mathematics*. The Clarendon Press, 1941.

5 J. R. Carson. Generalization of the reciprocal theorem. *Bell Labs Technical Journal*, 3(3):393–399, 1924.

6 J. R. Carson. Reciprocal theorems in radio communication. *Proceedings of the Institute of Radio Engineers*, 17(6):952–956, 1929.

7 P. M. Chirlian. *Basic Network Theory*. New York, NY: McGraw-Hill, 1969.

8 M. Cohen. Application of the reaction concept to scattering problems. *IRE Transactions on Antennas and Propagation*, 3(4):193–199, 1955.

9 R. E. Collin. *Antennas and Radiowave Propagation*. New York, NY: McGraw-Hill, 1985.

10 R. E. Collin. Limitations of the Thevenin and Norton equivalent circuits for a receiving antenna. *IEEE Antennas and Propagation Magazine*, 45(2):119–124, 2003.

11 A. T. de Hoop. A reciprocity relation between the transmitting and the receiving properties of an antenna. *Applied Scientific Research*, 19(1):90–96, 1968.

12 A. T. de Hoop. The *N*-port receiving antenna and its equivalent electrical network. *Philips Research Reports (issue in honour of C. J. Bouwkamp)*, 30:302–315, 1975.

13 A. T. de Hoop. A time domain energy theorem for scattering of plane electromagnetic waves. *Radio Science*, 19(5):1179–1184, 1984.

14 A. T. de Hoop. Time-domain reciprocity theorems for electromagnetic fields in dispersive media. *Radio Science*, 22(7):1171–1178, 1987.

15 A. T. de Hoop. Reciprocity, discretization, and the numerical solution of direct and inverse electromagnetic radiation and scattering problems. *Proceedings IEEE*, 79(10):1421–1430, 1991.

16 A. T. de Hoop. *Handbook of Radiation and Scattering of Waves*. London, UK: Academic Press, 1995.

17 A. T. de Hoop. A time-domain uniqueness theorem for electromagnetic wavefield modeling in dispersive, anisotropic media. *Radio Science Bulletin*, 305: 17–21, 2003.

18 A. T. de Hoop and G. de Jong. Power reciprocity in antenna theory. *Proceedings of the Institution of Electrical Engineers*, 121(10):594–605, 1974.

19 A. T. de Hoop, I. E. Lager, and V. Tomassetti. The pulsed-field multiport antenna system reciprocity relation and its applications: a time-domain approach. *IEEE Transactions on Antennas and Propagation*, 57(3):594–605, 2009.

20 G. Doetsch. *Introduction to the Theory and Application of the Laplace Transformation*. Berlin: Springer, 1974.

21 R. S. Elliott. *Antenna Theory and Design*. New York, NY: John Wiley & Sons, Inc., revised edition, 2003.

22 J. T. Fokkema and P. M. van den Berg. *Seismic Applications of Acoustic Reciprocity*. Elsevier, 1993.

23 J. T. Fokkema and P. M. van den Berg. 4D geophysical monitoring as an application of the reciprocity theorem. In P. M. van den Berg, H. Blok, and J. T. Fokkema, editors, *Wavefields and Reciprocity: Proceedings of a Symposium held in honour of Professor dr. A. T. de Hoop Delft, the Netherlands*, November 1996, pp. 99–108.

24 H. E. Green. Derivation of the Norton surface wave using the compensation theorem. *IEEE Antennas and Propagation Magazine*, 49(6):47–57, 2007.

25 R. C. Hansen. Relationships between antennas as scatterers and as radiators. *Proceedings of the IEEE*, 77(5):659–662, 1989.

26 R. F. Harrington. Matrix methods for field problems. *Proceedings of the IEEE*, 55(2):136–149, 1967.

27 R. F. Harrington. *Time-Harmonic Electromagnetic Fields*. Piscataway, NJ: IEEE Press, 2001.

28 R. F. Harrington and J. Mautz. Control of radar scattering by reactive loading. *IEEE Transactions on Antennas and Propagation*, 20(4):446–454, 1972.

29 A. Ishimaru. *Electromagnetic Wave Propagation, Radiation and Scattering*. Englewood Cliffs, NJ: Prentice-Hall, Inc., 1991.

30 J. A. Kong. *Electromagnetic Wave Theory*. New York, NY: John Wiley & Sons, Inc., 1986.

31 M. Lerch. Sur un point de la theorie des fonctions generatrices d'Abel. *Acta Mathematica*, 27:339–351, 1903.

32 Rayleigh Lord. *The Theory of Sound*, Vol. II. New York: Dover Publications, 1945.

33 H. A. Lorentz. The theorem of Poynting concerning the energy in the electromagnetic field and two general propositions concerning the propagation of light. *Versl. Kon. Akad. Wetensch. Amsterdam*, 4:176, 1896.

34 R. Mittra. A vector form of compensation theorem and its application to boundary-value problems. *Applied Scientific Research*, 11(1-2):26–42, 1964.

35 G. D. Monteath. Application of the compensation theorem to certain radiation and propagation problems. *Proceedings of the IEE–Part IV: Institution Monographs*, 98(1):23–30, 1951.

36 M. Rubinstein. An approximate formula for the calculation of the horizontal electric field from lightning at close, intermediate, and long range. *IEEE Transactions on Electromagnetic Compatibility*, 38(3):531–535, 1996.

37 V. H. Rumsey. Reaction concept in electromagnetic theory. *Physical Review*, 94(6):1483–1491, 1954.

38 S. Silver. *Microwave Antenna Theory and Design*. Number 19. New York, NY: McGraw-Hill, 1949.

39 J. A. Stratton. *Electromagnetic Theory*. New York: McGraw-Hill, 1941.

40 M. Stumpf. Controlling pulsed EM scattering of receiving antennas: the one-port case. In *Proceedings of the 2016 URSI International Symposium on Electromagnetic Theory*, pp. 211–214, 2016.

41 M. Stumpf. Limitations of the Cooray–Rubinstein formula: a time-domain analysis based on the Cagniard–DeHoop technique. *IEEE Transactions on Electromagnetic Compatibility*, 58(3):923–926, 2016.

42 M. Stumpf. Receiving-antenna Kirchhoff-equivalent circuits and their scattering reciprocity properties. *IET Microwaves, Antennas & Propagation*, 10(9):983–990, 2016.

43 M. Stumpf. A reciprocity relation of the time-correlation type and its application to antenna matching. *IEEE Transactions on Antennas and Propagation*, 64(5):1989–1993, 2016.

44 M. Stumpf. A generalization of the time-domain Cooray–Rubinstein formula. *IEEE Transactions on Electromagnetic Compatibility*, 59(5): 1638–1641, 2017.

45 W. L. Stutzman and G. A. Thiele. *Antennas Theory and Design*, 3rd ed. New York, NY: John Wiley & Sons, Inc., 2012.

46 J. G. van Bladel. *Electromagnetic Fields*. Hoboken, NJ: John Wiley & Sons, Inc. 2nd ed., 2007.

47 B. van der Pol and H. Bremmer. *Operational Calculus Based on the Two-Sided Laplace Integral*. New York, NY: Chelsea Publishing Company, 1987.

48 M. Stumpf and I. E. Lager. The time-domain optical theorem in antenna theory. *IEEE Antennas and Wireless Propagation Letters*, 14:895–897, 2015.

49 J. R. Wait. Concerning the horizontal electric field of lightning. *IEEE Transactions on Electromagnetic Compatibility*, 39(2):186, 1997.

50 W. Welch. Reciprocity theorems for electromagnetic fields whose time dependence is arbitrary. *IRE Transactions on Antennas and Propagation*, 8(1):68–73, 1960.

51 W. Welch. Comments on "Reciprocity theorems for electromagnetic fields whose time dependence is arbitrary". *IRE Transactions on Antennas and Propagation*, 9(1):114–115, 1961.

52 D. V. Widder. *The Laplace Transform*. Princeton, NJ: Princeton University Press, 1946.

INDEX

Electromagnetic Reciprocity in Antenna Theory, First Edition. Martin Stumpf.